11

CRM
SERIES

Centro
di Ricerca
Matematica
Ennio De Giorgi

Franck Barthe
Institut de Mathématiques de Toulouse
Université Paul Sabatier
31062 Toulouse Cedex 9, France
barthe@math.univ-toulouse.fr

Wilfrid Gangbo
School of Mathematics
Georgia Institute of Technology
Atlanta, GA 30332, USA
gangbo@math.gatech.edu

Francesco Maggi
Dipartimento di Matematica "U. Dini"
Università degli Studi di Firenze
Viale Morgagni 67/A
50134, Firenze, Italia
maggi@math.unifi.it

Adrian Tudorascu
Department of Mathematics
University of Wisconsin
Madison, WI 53706, USA
tudorasc@math.wisc.edu

Optimal Transportation, Geometry and Functional Inequalities

edited by
Luigi Ambrosio

EDIZIONI
DELLA
NORMALE

ISBN: 978-88-7642-373-4

Contents

Preface

Starting from 2000, a meeting on the theory of optimal mass transportation and its applications has been held at the Scuola Normale. In 2008, on the occasion of the fourth edition of this meeting, I decided with my fellow organizers (Giuseppe Buttazzo, Giuseppe Savaré and Nicola Fusco) to add a school on this topic, partially funded by GNAMPA and PRIN.

The lectures delivered by Franck Barthe, Wilfrid Gangbo, Francesco Maggi and Robert McCann were so nice and rich of recent developments that we thought of turning them into Lecture Notes. The first three lecturers agreed to this project, and with the additional help of Alessio Figalli, Alessandro Carlotto and Luigi De Pascale all notes have been typed, collected, typed and revised.

The book presents in a friendly and unitary way many recent developments of the theory which have not previously appeared in book form. Topics include concentration inequalities, evolution problems of Hamiltonian type and sharp isoperimetric inequalities.

Pisa, February 5, 2010

Luigi Ambrosio

Transportation techniques and Gaussian inequalities

Franck Barthe

Abstract. These lecture notes illustrate how optimal transportation techniques may apply to the study of Gaussian measures and related probability distributions. The focus is on dimension free concentration inequalities and correlation of convex sets.

1 Introduction

Optimal transportation has attracted a lot of attention in recent years and has proved applicable to various and sometimes unexpected fields. We refer to the books [1,40,47,48] for more details and references. In these notes, we present applications of transportation techniques to the study of the Gaussian measure, which is of central importance in probability theory, and of related distributions. Our specific goal is to expose recent progress on the understanding of dimension-free Gaussian-type concentration and on the famous Gaussian correlation conjecture. More precisely, Section 2 presents a complete picture of the relationships between Gaussian concentration, quadratic transportation cost inequalities and logarithmic Sobolev inequalities, which was recently completed by contributions of Gozlan [19] and Cattiaux and Guillin [12]. Section 3 presents two striking results of Hargé [21, 22] which confirm in particular cases the Gaussian correlation conjecture for symmetric convex sets. The material of these lecture notes comes from research articles and was not yet covered by the textbooks in the field. We have done our best to give a self-contained exposition, which has lead us to reproduce also some classical results and their proofs. Next we set some notation used throughout these notes:

- Euclidean space $\mathbb{R}^n$ with norm $|\cdot|$ and scalar product $\langle \cdot, \cdot \rangle$;
- $\mathscr{L}^n$ Lebesgue measure in $\mathbb{R}^n$;
- γ_n standard Gaussian in $\mathbb{R}^n$, with density $(2\pi)^{-n/2} \exp(-|x|^2/2)$;

- B^n the open unit ball in $\mathbb{R}^n$ centered at the origin;
- $\|F\|_{\text{Lip}}$ the (smallest) Lipschitz constant of a function F;
- $\mathbb{E}_\mu[F]$ expectation of a random variable F with respect to μ, or $\mathbb{E}[F]$ when there is no ambiguity on μ;
- $\mathscr{P}(X)$ space of Borel probability measures in X;
- $\mathscr{B}(X)$ σ-algebra of Borel sets of X;
- χ_E characteristic function of E;
- given two sets A, $B \subset \mathbb{R}^n$, their (Minkowski) sum is $A + B = \{a + b : (a, b) \in A \times B\}$.

ACKNOWLEDGEMENTS. It is a pleasure to conclude this introduction by thanking the organizers of the GNAMPA school for all their work and Luigi Ambrosio, Alessandro Carlotto and Alessio Figalli for their enormous help in the preparation of these notes.

2 Gaussian concentration

2.1 Basic definitions and an application

What is Gaussian concentration? This property can be stated in a strong from (discovered independently by Sudakov and Tsirel'son [43] and Borell [9]) and a weak form, the latter being more suitable for extensions to non-Gaussian measures.

Strong form ([9,43])

Let $r > 0$. If $A \in \mathscr{B}(\mathbb{R}^n)$ and H is a halfspace satisfying $\gamma_n(A) = \gamma_n(H)$, then denoting by $A_r := A + rB^n$ the r-enlargement of A, we have

$$\gamma_n(A_r) \geq \gamma_n(H_r).$$

Weak form

If $A \in \mathscr{B}(\mathbb{R}^n)$ and $\gamma_n(A) \geq 1/2$, then

$$\gamma_n(A_r) \geq 1 - e^{-r^2/2} \qquad \forall r \geq 0. \tag{2.1}$$

The strong form implies the weak form because $\gamma_n(H_r) \geq 1 - e^{-r^2/2}$ whenever H is a halfspace with $\gamma_n(H) \geq 1/2$. The weak formulation can also be given in functional terms as follows: let $F : \mathbb{R}^n \to \mathbb{R}$ be a

Lipschitz map, let m be a median of F relative to γ_n (*i.e.* $\gamma_n(\{F \geq m\}) \geq 1/2$ and $\gamma_n(\{F \leq m\}) \geq 1/2$) and apply (2.1) with $A = \{F \leq m\}$ to get

$$1 - e^{-r^2/2} \leq \gamma_n(\{F \leq m\}_r) \leq \gamma_n(\{F \leq m + r\|F\|_{\mathrm{Lip}}\}),$$

so that, setting $t = r\|F\|_{\mathrm{Lip}}$ gives $\gamma_n(\{F > m + t\}) \leq e^{-t^2/2\|F\|_{\mathrm{Lip}}^2}$. The same argument, applied to $-F$, gives eventually

$$\gamma_n(\{|F - m| > t\}) \leq 2e^{-t^2/2\|F\|_{\mathrm{Lip}}^2}. \tag{2.2}$$

Roughly speaking, this means that F is very concentrated around its median. Integrating (2.2) in $(0, +\infty)$ yields

$$\int |F - m|^p d\gamma_n = \int_0^\infty \gamma_n(\{|F - m|^p > t\}) dt$$
$$\leq p2^{p/2}\Gamma\left(\frac{p}{2}\right)\|F\|_{\mathrm{Lip}}^p. \tag{2.3}$$

For $p = 1$, the above inequality implies that

$$|\mathbb{E}_{\gamma_n}[F] - m| \leq c\|F\|_{\mathrm{Lip}} \tag{2.4}$$

for some numerical constant c. Hence one may easily replace m by the expectation $\mathbb{E}_{\gamma_n}[F]$ in (2.2), getting the inequality

$$\gamma_n(\{|F - \mathbb{E}_{\gamma_n}[F]| > t\}) \leq 2e^{-(t-c\|F\|_{\mathrm{Lip}})^2/2\|F\|_{\mathrm{Lip}}^2} \quad \forall t \geq c\|F\|_{\mathrm{Lip}}. \tag{2.5}$$

The restriction on t can be removed at the price of worse numerical constants. Indeed the latter inequality implies that

$$\gamma_n(\{|F - \mathbb{E}_{\gamma_n}[F]| > t\}) \leq \max\left(2, e^{c^2/2}\right)e^{-t^2/8\|F\|_{\mathrm{Lip}}^2} \quad \forall t \geq 0,$$

which is trivial if $t \leq 2c\|F\|_{\mathrm{Lip}}$. Also, note for further use that combining the properties of the variance with (2.3) for $p = 2$ gives

$$0 \leq \mathbb{E}_{\gamma_n}\left[F^2\right] - \mathbb{E}_{\gamma_n}[F]^2 = \mathbb{E}_{\gamma_n}\left[(F - \mathbb{E}_{\gamma_n}[F])^2\right]$$
$$\leq \mathbb{E}_{\gamma_n}\left[(F - m)^2\right] \leq 4\|F\|_{\mathrm{Lip}}^2. \tag{2.6}$$

The concentration phenomenon, for the Gaussian and other measures, has many applications in probability theory, see *e.g.* [27]. Its relevance was first put forward by V. Milman in the setting of Banach space geometry, for random constructions of remarkable objects, see [34]. As a motivation, let us present an example of a striking application. The Johnson-Lindenstrauss flattening Lemma [23], an easy consequence of concentration, is of fundamental importance in metric geometry and theoretical computer science (see *e.g.* [33] and the references therein).

Lemma 2.1 (Johnson-Lindenstrauss [23]). *Let $p_1, \ldots, p_N$ be points in the Euclidean space $\mathbb{R}^n$. Let $\varepsilon > 0$. Then there exists a linear map $\pi : \mathbb{R}^n \to \mathbb{R}^k$ such that*

$$(1-\varepsilon)|p_i - p_j| \leq |\pi(p_i) - \pi(p_j)| \leq (1+\varepsilon)|p_i - p_j| \qquad 1 \leq i, j \leq N,$$

provided $k \geq c\varepsilon^{-2} \ln N$, with c universal constant. Such a map π is called, for obvious reasons, a "quasi isometry". Note that, surprisingly, $k \sim \ln N \ll N$.

Since the result does not depend on the initial dimension, the lemma readily extends to N points in a Hilbert space, and provides a linear map into a k-dimensional subspace, which is quasi-isometric on these points.

In order to prove the lemma, let us consider independent standard Gaussian variables g_{ij}, $1 \leq i \leq k$, $1 \leq j \leq n$; these maps can be realized on the probability space $\mathbb{R}^{kn}$ endowed with γ_{kn}, so that g_{ij} is the ij-th entry of a "random" matrix G. For $x \in \mathbb{R}^n$ fixed, let us consider the map $G \mapsto G(x)$ as a random variable in $\mathbb{R}^{kn}$. The map $G \mapsto |Gx|$ is obviously $|x|$-Lipschitz and

$$\int |Gx|^2 \, d\gamma_{kn}(G) = k|x|^2.$$

If $x \in S^{n-1}$, by (2.4) and (2.6) the median m_k of $G \mapsto |Gx|$, the expectation $\mathbb{E}[|Gx|]$ and the standard deviation $\sqrt{\mathbb{E}[|Gx|^2]}$ have all the same behavior, *i.e.* $\sqrt{k}$, as $k \to \infty$; also, rotation invariance implies that m_k is independent of x.

By applying the concentration property we get

$$\gamma_{kn}\left(\{G : ||Gx| - m_k|x|| > t|x|\}\right) \leq 2e^{-t^2/2}$$

and, if we set $t = \varepsilon m_k$,

$$\gamma_{kn}\left(\{G : |Gx| \notin [(1 - \varepsilon)m_k|x|, (1 + \varepsilon)m_k|x|]\}\right) \leq 2e^{-\varepsilon^2 m_k^2/2}.$$

Repeating this procedure for ℓ points $x^1, \ldots, x^\ell \in \mathbb{R}^n$ we get

$$\gamma_{kn}\left(\{G : |Gx^i| \notin [(1 - \varepsilon)m_k|x^i|, (1 + \varepsilon)m_k|x^i|] \text{ for some } i\}\right)$$
$$\leq 2\ell e^{-\varepsilon^2 m_k^2/2} < 1$$

provided $m_k^2 \geq 2\ln(2\ell)/\varepsilon^2$, and this surely holds if $c^2 k \geq 2\ln(2\ell)/\varepsilon^2$, choosing c universal constant such that $m_k \geq c\sqrt{k}$. This means that there is at least one G such that

$$(1 - \varepsilon)|x^i| \leq \frac{1}{m_k}|Gx^i| \leq (1 + \varepsilon)|x^i| \qquad \forall i = 1, \ldots, \ell.$$

To conclude, it suffices to apply this construction with $\ell = N(N-1)/2$ and x equal to $p_i - p_j$, $1 \leq i, j \leq N$. $\qquad\square$

It is natural to look for probability distributions having the same concentration as the standard Gaussian measure. This motivates the following

Definition 2.2. If μ is a probability measure on $\mathbb{R}^d$, we say that μ has a Gaussian concentration (with constants $c, C > 0$) if, for all 1-Lipschitz $F : \mathbb{R}^d \to \mathbb{R}$ and all medians m of F, it holds

$$\mu(\{|F - m| > t\}) \leq C e^{-ct^2} \qquad \forall t \geq 0 \qquad (2.7)$$

for some constants $c, C > 0$. Moreover, we say that μ enjoys a dimension-free Gaussian concentration if there are constants $c, C > 0$ not depending on $n \geq 1$ such that for all $n \geq 1$, μ^n has Gaussian concentration with constants c, C.

Obviously (2.7) provides interesting information for large values of t; an equivalent formulation is then the existence of constant $C' \geq 0, c' > 0$ and $t_0 \geq 0$ satisfying

$$\mu^n(\{|F - m| > t\}) \leq C' e^{-c'(t-t_0)^2} \qquad \forall t \geq t_0.$$

By the same argument relating (2.2), (2.4) and (2.5), an equivalent formulation could be gives using $\mathbb{E}_{\mu^n}[F]$ instead of the median.

In the previous definition $\mu^n = \mu \times \cdots \times \mu$ denotes the product of n copies of μ. Let us emphasize that dimension-free Gaussian concentration allows to infer strong distributional properties of sequences of independent random variables with distribution μ. There exist by now various techniques to derive concentration inequalities for various product measures. Among the most successful are infimum convolution inequalities, (quadratic) transportation-cost inequalities (T_2 for short) or logarithmic Sobolev inequalities (LS for short). We will give more precise definitions and references later on. The goal of this section is to present a complete proof of the relationships between the later two and concentration. The situation can be summed up schematically as follows:

$$LS \implies T_2 \iff \text{dimension-free Gaussian concentration}$$

and $T_2 \not\Longrightarrow LS$.

2.2 Logarithmic Sobolev inequalities

Now we analyze the relations between Gaussian concentration and logarithmic Sobolev inequalities. We refer to the survey [26] by Ledoux for an extensive presentation of the log-Sobolev approach to measure concentration. Let us also emphasize that log-Sobolev inequalities have other important consequences, as hypercontractive estimates or exponential decay of entropy for some evolution equations, see *e.g.* [2].

Definition 2.3 (Log-Sobolev inequality). If $\mu \in \mathscr{P}(\mathbb{R}^d)$ with $\mu \ll \mathscr{L}^d$, we say that μ satisfies a Log-Sobolev inequality with constant C, and write $\mu \in LS(C)$, if

$$\mathrm{Ent}_\mu(f^2) \leq 2C \int |\nabla f|^2 \, d\mu$$

holds for any locally Lipschitz function $f : \mathbb{R}^d \to \mathbb{R}$.

Here $\mathrm{Ent}_\mu(g)$ is the *relative entropy* functional with respect to μ, defined for all g with $g \ln g \in L^1(\mu)$ by

$$\mathrm{Ent}_\mu(g) := \int g \ln g \, d\mu - \left(\int g \, d\mu \right) \ln \left(\int g \, d\mu \right). \tag{2.8}$$

Notice that $\mathrm{Ent}_\mu(g)$ is defined in such a way that $\mathrm{Ent}_\mu(\lambda g) = \lambda \mathrm{Ent}_\mu(g)$ for all $\lambda \geq 0$; when g is a probability density (*i.e.* $\int g \, d\mu = 1$) it reduces simply to $\int g \ln g \, d\mu$. We shall also use the notation $\mathrm{Ent}(\nu|\mu)$ for the *relative entropy* of ν with respect to μ:

$$\mathrm{Ent}(\nu|\mu) := \begin{cases} \mathrm{Ent}_\mu(f) & \text{if } \nu = f\mu \text{ and } f \ln f \in L^1(\mu); \\ +\infty & \text{otherwise.} \end{cases}$$

Notice also that $\mathrm{Ent}(\nu|\mu)$ is defined for all ν and μ. We collect in the Appendix more properties of $\mathrm{Ent}(\nu|\mu)$.

The inequality (2.8) above makes sense thanks to Rademacher's theorem, which provides the existence of ∇f $\mathscr{L}^d$-a.e. and then μ-a.e., by the absolute continuity assumption. Since Ent_μ is not defined on $L^1(\mu)$, the Log-Sobolev inequality should be interpreted as follows: whenever the right hand side is finite, then $f^2 \ln f^2 \in L^1(\mu)$ and the inequality holds.

A remarkable property of the Log-Sobolev inequality is its stability under tensor products.

Proposition 2.4. *Let* $\mu \in \mathscr{P}(\mathbb{R}^d)$. *Then* $\mu \in LS(C)$ *implies* $\mu^n \in LS(C)$.

One can prove more, namely $\mu \in LS(C)$ and $\nu \in LS(C)$ implies $\mu \times \nu \in LS(C)$. This implication is a direct consequence of the additivity of entropy (4.7), proved in the appendix, taking into account that

$$\mathrm{Ent}_{\mu}\left(\int f^2(x, \cdot)\, dv\right) \leq 2C \int \left| \nabla \sqrt{\int f^2(x, \cdot)\, dv} \right|^2 d\mu$$

$$= 2C \int \frac{|\int f(x, \cdot) \nabla_x f(x, \cdot)\, dv|^2}{\int f^2(x, \cdot)\, dv}\, d\mu$$

$$\leq 2C \int |\nabla_x f|^2\, dv\, d\mu$$

and

$$\int \mathrm{Ent}_v(f^2(x, \cdot))\, d\mu \leq 2C \int \int |\nabla_y f|^2\, dv\, d\mu.$$

Next, we present Herbst's argument proving the implication

$$LS(C) \Longrightarrow \text{Gaussian concentration.}$$

Let $v \in \mathcal{P}(\mathbb{R}^d)$ such that $v \in LS(C)$. First of all, we notice that we can reduce ourselves to check (2.7) for bounded 1-Lipschitz functions $F : \mathbb{R}^d \to \mathbb{R}$, by a truncation argument (see e.g. [26] for technical details). Then, for $\lambda \in \mathbb{R}$ we consider

$$H(\lambda) := \int_{\mathbb{R}^d} e^{\lambda F}\, dv$$

whose derivative satisfies $\lambda H'(\lambda) = \int \lambda F e^{\lambda F}\, dv$. Then, we can apply the log-Sobolev inequality for $f^2 = e^{\lambda F}$ relative to the measure v to obtain

$$\lambda H'(\lambda) - H(\lambda) \ln H(\lambda) \leq 2C \left(\frac{\lambda}{2}\right)^2 \int |\nabla F|^2 e^{\lambda F}\, dv.$$

Taking into account that F is 1-Lipschitz we get

$$\lambda H'(\lambda) - H(\lambda) \ln H(\lambda) \leq C\frac{\lambda^2}{2} H(\lambda),$$

so that $\frac{d}{d\lambda}(\ln H(\lambda))/\lambda \leq C/2$. Now, a simple Taylor expansion shows that the function $(\ln H(\lambda))/\lambda$ converges to $\int F\, dv$ as $\lambda \downarrow 0$. Consequently, an integration gives

$$H(\lambda) \leq e^{\lambda \mathbb{E}_v[F] + C\lambda^2/2}.$$

Finally we use the Markov inequality for $\lambda > 0$

$$v(\{F - \mathbb{E}_v[F] > t\}) = v(\{e^{\lambda(F - \mathbb{E}_v[F])} > e^{\lambda t}\})$$

$$\leq \frac{1}{e^{\lambda t}} \int e^{\lambda F - \lambda \mathbb{E}_v[F]}\, dv \leq e^{-\lambda t + C\lambda^2/2}$$

and for $\lambda = t/C$ we get

$$\nu\big(\{|F - \mathbb{E}_\nu[F]| > t\}\big) \leq 2e^{-t^2/2C},$$

from which a concentration inequality with respect to the median, of the form (2.7), easily follows.

Combining Herbst's argument with the tensorisation property stated in Proposition 2.4 yields

$$\mu \in LS(C) \Longrightarrow \forall n, \ \mu^n \in LS(C) \Longrightarrow \mu$$

has dimension-free Gaussian concentration.

2.3 Transportation cost inequalities

Let us first recall some basic definitions and properties relative to the optimal transport problem; a general reference for all the properties here stated without proof is [47].

Definition 2.5 (p-th transportation cost and Wasserstein distance).
For $p \geq 1$ and $\nu_1, \nu_2 \in \mathscr{P}(\mathbb{R}^d)$ we set

$$T_p(\nu_1, \nu_2) := \inf_\pi \int_{\mathbb{R}^d \times \mathbb{R}^d} |x - y|^p \, d\pi(x, y)$$

where the infimum runs among all $\pi \in \mathscr{P}(\mathbb{R}^d \times \mathbb{R}^d)$ having respectively ν_1 and ν_2 as marginals (i.e. $\pi(A \times \mathbb{R}^d) = \nu_1(A)$ and $\pi(\mathbb{R}^d \times B) = \nu_2(B)$ for all $A, B \in \mathscr{B}(\mathbb{R}^d)$).

We denote also by W_p the p-Wasserstein distance:

$$W_p(\nu_1, \nu_2) := \big[T_p(\nu_1, \nu_2)\big]^{1/p}.$$

Throughout these notes we shall denote by $\Gamma(\nu_1, \nu_2) \subset \mathscr{P}(\mathbb{R}^d \times \mathbb{R}^d)$ the space of all π having ν_1 and ν_2 has marginals, and by $\Gamma_0(\nu_1, \nu_2)$ those for which the infimum is attained. The set $\Gamma_0(\nu_1, \nu_2)$ is non-empty when ν_1, ν_2 belong to

$$\mathscr{P}_p(\mathbb{R}^d) = \left\{\mu \in \mathscr{P}(\mathbb{R}^d) : \int |x|^p \, d\mu(x) < +\infty\right\}.$$

It turns out that $\mathscr{P}_p(\mathbb{R}^d)$ is a complete and separable metric space, when endowed with the W_p distance (finiteness of moments ensures finiteness of distance). In addition $W_p(\mu_n, \mu) \to 0$ is equivalent to

$$\int |x|^p \, d\mu_n(x) \to \int |x|^p \, d\mu(x) \text{ and}$$

$$\int \phi \, d\mu_n \to \int \phi \, d\mu, \quad \forall \phi \in C_c(\mathbb{R}^d). \tag{2.9}$$

The definitions of T_p and W_p obviously extend to more general spaces. We shall also need a duality formula for the 2-transportation cost:

$$T_2(\mu, \nu)$$
$$= \sup \left\{ \int f \, d\mu + \int g \, d\nu : \forall x, y \in \mathbb{R}^d, \ f(x) + g(y) \leq |x - y|^2 \right\}. \qquad (2.10)$$

Notice that, by Jensen's inequality, $\mathrm{Ent}(\nu|\mu) \geq 0$. In addition, since $t \mapsto t \ln t$ is strictly convex, it is not hard to check that equality holds if and only if $\nu = \mu$. For this reason $\mathrm{Ent}(\nu|\mu)$ can be considered as a "distance" of ν from μ. This motivates the next definition.

Definition 2.6. Let $\mu \in \mathscr{P}(\mathbb{R}^d)$. We say that $\mu \in T_p(C)$ if

$$W_p(\nu, \mu) \leq \sqrt{C \, \mathrm{Ent}(\nu|\mu)} \qquad \forall \nu \in \mathscr{P}(\mathbb{R}^d).$$

Such transportation cost inequalities were introduced (for $p = 1$) by Marton [32]. Now we present her argument showing that

$$T_1(C) \Longrightarrow \text{Gaussian concentration.}$$

Let $A \in \mathscr{B}(\mathbb{R}^d)$, let $r > 0$ and let $A' = \mathbb{R}^d \setminus A_r$. If we introduce

$$\mu_A(B) := \frac{\mu(A \cap B)}{\mu(A)}, \qquad \mu_{A'}(B) := \frac{\mu(A' \cap B)}{\mu(A')}$$

the normalized restrictions of μ to A and A' respectively, it is obvious that $W_1(\mu_A, \mu_{A'}) \geq r$, because $|x - y| \geq r$ for all $x \in A$ and $y \in A'$. The triangle inequality gives

$$W_1(\mu_A, \mu_{A'}) \leq W_1(\mu_A, \mu) + W_1(\mu, \mu_{A'})$$

$$\leq \sqrt{C \mathrm{Ent}(\mu_A|\mu)} + \sqrt{C \mathrm{Ent}(\mu_{A'}|\mu)}$$

$$= \sqrt{C \ln \frac{1}{\mu(A)}} + \sqrt{C \ln \frac{1}{1 - \mu(A_r)}}.$$

Rearranging the terms we get

$$\mu(A_r) \geq 1 - e^{-\left(r - \sqrt{C \ln 1/\mu(A)}\right)^2 / C}$$

if $r \geq \sqrt{C \ln 1/\mu(A)}$. This obviously implies the weak form of Gaussian concentration and (2.7). $\qquad \square$

Unlike the log-Sobolev inequality, the T_1 inequality does not enjoy the tensorisation property. This has motivated the introduction by Talagrand [44] of the T_2 inequality, which does:

Theorem 2.7 (Talagrand [44]). $\mu \in T_2(C)$ *implies* $\mu^n \in T_2(C)$.

More generally, the same result with the same proof applies to products of probability measures μ in arbitrary metric spaces, if we choose the "Hilbertian" distance in the product space to define W_2. As a consequence, the inequality $W_1 \leq W_2$ gives

$$\mu \in T_2(C) \Longrightarrow \forall n, \ \mu^n \in T_2(C) \Longrightarrow \forall n, \ \mu^n \in T_1(C)$$

$$\Longrightarrow \begin{array}{l} \mu \text{ has dimension-free} \\ \text{Gaussian concentration.} \end{array}$$

Let us now prove Talagrand's theorem. We shall actually prove more, namely that $\mu \in T_2(C)$ and $\nu \in T_2(C)$ implies $\mu \times \nu \in T_2(C)$. We shall denote respectively by (X, d_X) and (Y, d_Y) the spaces where the measures μ and ν live. Let $\theta \in \mathscr{P}(X \times Y)$ and let us represent it as $d\theta_x(y)d\sigma(x)$, where σ is the first marginal of θ and $\theta_x \in \mathscr{P}(Y)$: this means that we have the disintegration formula

$$\int_{X \times Y} \phi \, d\theta = \int_X \left(\int_Y \phi(x, y) \, d\theta_x(y) \right) d\sigma(x)$$

for all bounded Borel functions $\phi(x, y)$. We want to compare this measure to $\mu \times \nu$. The heuristic idea is to rearrange mass first in the x variable, considering the marginals σ and μ, and then in the y variable, considering the marginals θ_x and ν. To make this idea precise, we choose $p \in \Gamma_0(\mu, \sigma) \subset \mathscr{P}(X \times X)$ and $\pi_x \in \Gamma_0(\nu, \theta_x) \subset \mathscr{P}(Y \times Y)$ and define

$$d\pi(x_1, y_1, x_2, y_2) := d\pi_{x_2}(y_1, y_2)dp(x_1, x_2) \in \mathscr{P}((X \times Y) \times (X \times Y)).$$

A technical point: thanks to the measurability of $x \mapsto \theta_x$ there is a measurable selection of optimal plans $x \mapsto \pi_x$, so that π makes sense. Then, the rest of the proof is simply a matter of computations: first of all we check that the second marginal of π, relative to (x_2, y_2), is θ (the fact that the first one, relative to (x_1, y_1), is $\mu \times \nu$ is obvious):

$$\int_{(X \times Y)^2} \phi(x_2, y_2) \, d\pi = \int_{X^2} \int_{Y^2} \phi(x_2, y_2) \, d\pi_{x_2}(y_1, y_2) \, dp(x_1, x_2)$$

$$= \int_X \int_Y \phi(x, y) \, d\theta_x(y) \, d\sigma(x)$$

$$= \int_{X \times Y} \phi(x, y) \, d\theta(x, y).$$

Now we compute the cost (we omit for simplicity of notation the domain of integration):

$$T_2(\mu \times, \nu, \theta) \le \int \left(|x_1 - x_2|_X^2 + |y_1 - y_2|_Y^2 \right) d\pi$$

$$= \int |x_1 - x_2|_X^2 \, dp + \int \int |y_2 - y_1|_Y^2 \, d\pi_{x_2} dp$$

$$= T_2(\mu, \sigma) + \int T_2(\nu, \theta_x) \, d\sigma(x)$$

$$\le C \text{Ent}(\sigma \,|\, \mu) + C \int \text{Ent}(\theta_x \,|\, \nu) \, d\sigma(x) = C \text{Ent}(\theta \,|\, \mu \times \nu).$$

In the last equality we use (4.3), see the appendix.   □

2.4 T_2 and large deviations

This subsection is devoted to a recent result of Gozlan [19], providing the implication

$$\text{dimension-free Gaussian concentration} \implies T_2.$$

The precise statement is:

Theorem 2.8 (Gozlan [19]). *Let μ be a probability measure in $\mathbb{R}^d$. If there exist r_0, $b \ge 0$ and $a > 0$ such that*

$$\mu^n(A) \ge 1/2 \implies \mu^n(A_r) \ge 1 - b e^{-a(r-r_0)^2} \quad \forall r \ge r_0 \qquad (2.11)$$

then $\mu \in T_2(1/a)$.

In order to prove this theorem we need some auxiliary results from probability theory of an independent interest. Recall that the *law of large numbers* says that if (X_i) is a sequence of i.i.d. (independent and identically distributed) random variables in an abstract probability space $(\Omega, \mathbb{P})$, with $\mathbb{E}[|X_i|^p] < \infty$ for some $p \in [1, +\infty)$, then the arithmetic means

$$U_n := \frac{X_1 + \ldots + X_n}{n}$$

converge almost surely and in L^p to $\mathbb{E}[X_i]$. Many classical tools are available to estimate the rate of convergence, for instance the central limit theorem. Here we recall the following classical result which is the starting point of the theory of large deviations (see *e.g.* [15, 16]).

Theorem 2.9 (Cramer). *Let (X_i) be real valued, i.i.d. and satisfying $\mathbb{E}[e^{tX_i}] < \infty$ for all $t \in (-\varepsilon, \varepsilon)$ for some $\varepsilon > 0$. Set*

$$\Lambda^*(x) := \sup_{y \in \mathbb{R}} xy - \ln \mathbb{E}[e^{yX_i}] \geq 0 \qquad \forall x \in \mathbb{R}. \qquad (2.12)$$

Then, for all Borel sets $A \subset \mathbb{R}$

$$-\inf_{a \in \text{Int}(A)} \Lambda^*(a) \leq \liminf_{n \to \infty} \frac{1}{n} \ln \mathbb{P}\left(\{U_n \in A\}\right)$$

$$\leq \limsup_{n \to \infty} \frac{1}{n} \ln \mathbb{P}\left(\{U_n \in A\}\right) \leq -\inf_{a \in \bar{A}} \Lambda^*(a).$$

For shortness, one says that the sequence (U_n) satisfies a large deviation principle with speed n and rate function Λ^*. Roughly speaking, this result means that generically the event $\{U_n \in A\}$ has probability $\mathbb{P}(\{U_n \in A\}) \sim e^{-n \inf_A \Lambda^*}$. To make more precise the concept of "generic", notice that if you have a family of open sets A_t with $\overline{A}_t \subset A_s$ for $t > s$ (e.g. the sets $\{f > t\}$ with f continuous) then $\inf_{\overline{A}_t} \Lambda^* = \inf_{A_t} \Lambda^*$ with at most countably many exceptions.

We need also to consider empirical measures. To this aim, let $\mu \in \mathscr{P}(\mathbb{R}^d)$ with finite quadratic moments, let $(X_i)_{i \geq 1}$ be a sequence of independent random vectors of $\mathbb{R}^d$ with law μ (on an abstract probability space $(\Omega, \mathbb{P})$) and denote by

$$\mu_n := \frac{1}{n} \sum_{i=1}^{n} \delta_{X_i} \qquad (2.13)$$

the induced family of atomic random measures. We need the following basic facts:

(a) the map $x \in (\mathbb{R}^d)^n \mapsto W_2(\sum_1^n \frac{1}{n}\delta_{x_i}, \mu)$ is $1/\sqrt{n}$-Lipschitz;
(b) $W_2(\mu_n, \mu) \to 0$ almost surely as $n \to +\infty$;
(c) $\mu^n\left(\{x : W_2(\sum_1^n \frac{1}{n}\delta_{x_i}, \mu) > \delta\}\right) \to 0$ for all $\delta > 0$. In particular the median of $W_2(\sum_1^n \frac{1}{n}\delta_{x_i}, \mu)$ under μ^n tends to 0 as $n \to \infty$.

In order to show (a), notice that

$$\left| W_2\left(\sum_1^n \frac{1}{n}\delta_{x_i}, \mu\right) - W_2\left(\sum_1^n \frac{1}{n}\delta_{y_i}, \mu\right) \right| \leq W_2\left(\sum_1^n \frac{1}{n}\delta_{x_i}, \sum_1^n \frac{1}{n}\delta_{y_i}\right).$$

Now, the plan $\pi = \sum_1^n \frac{1}{n}\delta_{(x_i, y_i)} \in \Gamma\left(\sum_1^n \frac{1}{n}\delta_{x_i}, \sum_1^n \frac{1}{n}\delta_{y_i}\right)$ gives

$$T_2\left(\sum_1^n \frac{1}{n}\delta_{x_i}, \sum_1^n \frac{1}{n}\delta_{y_i}\right) \leq \frac{|x - y|^2}{n}.$$

In order to prove (b), we observe that the law of large numbers applied to the i.i.d. random variables $|X_i|^2$ gives

$$\int_{\mathbb{R}^d} |x|^2 \, d\mu_n = \frac{1}{n} \sum_{i=1}^n |X_i|^2 \to \mathbb{E}[|X_i|^2] = \int_{\mathbb{R}^d} |x|^2 \, d\mu$$

almost surely. A similar argument provides the almost sure convergence $\int \phi(x) \, d\mu_n \to \int \phi(x) \, d\mu$ for all $\phi \in C_c(\mathbb{R}^d)$. Choosing a countable dense set $D \subset C_c(\mathbb{R}^d)$ we can ensure, besides the almost sure moment convergence, the almost sure convergence for all $\phi \in D$. By (2.9) and a density argument we conclude that $W_2(\mu_n, \mu) \to 0$ almost surely.

Finally, (c) easily follows (b) by the identity

$$\mu^n \left(\left\{ x : W_2 \left(\sum_1^n \frac{1}{n} \delta_{x_i}, \mu \right) > \delta \right\} \right) = \mathbb{P} \left(\{ \omega : W_2(\mu_n(\omega), \mu) > \delta \} \right).$$

Considering (b) as a version of the law of large numbers for $\mathscr{P}(\mathbb{R}^d)$-valued random variables, it is natural to ask whether an analog of Cramer's theorem is available in this context. An affirmative answer is provided by Sanov's large deviations theorem (see *e.g.* [16, Theorem 6.2.10]): in this case the rate function is precisely the relative entropy functional $\nu \mapsto \mathrm{Ent}(\nu|\mu)$. Following the presentation of [19] and in order to avoid topological considerations, we restrict our attention to the sets $A = \{\nu : W_2(\nu, \mu) > t\}$ and to the lim inf inequality. The appropriate version of Sanov's theorem that we will need is stated next:

Theorem 2.10. *If $\mu \in \mathscr{P}(\mathbb{R}^d)$ has finite quadratic moments, then*

$$\liminf_{n \to \infty} \frac{1}{n} \ln \mu^n \left(\left\{ x : W_2 \left(\sum_{i=1}^n \frac{1}{n} \delta_{x_i}, \mu \right) > t \right\} \right)$$
$$\geq - \inf \left\{ \mathrm{Ent}(\nu|\mu) : W_2(\nu, \mu) > t \right\}. \tag{2.14}$$

The idea of the proof is as follows: if we equip $(\mathbb{R}^d)^n$ with the product measure ν^n and consider the coordinate projections $x \mapsto x_i$, $i = 1, \ldots, n$, then x_i are independent and identically distributed, with law ν. Then, in order to estimate "$\mu^n\{\sum_1^n \frac{1}{n}\delta_{x_i} \to \nu\}$" we consider the quantity "$\nu^n\{\sum_1^n \frac{1}{n}\delta_{x_i} \to \nu\}$", close to 1 by the law of large numbers, and estimate the change in the reference measure in terms of the relative entropy. The comparison is provided by the next classical lemma, which can be found in [15, page 76].

Lemma 2.11. *Let $A \in \mathcal{B}(\mathbb{R}^d)$ and $\nu \ll \mu$. Assume that the set $\{x : \sum_1^n \frac{1}{n}\delta_{x_i} \in A\}$ belongs to $\mathcal{B}(\mathbb{R}^{dn})$ and satisfies $\nu^n(\{x : \sum_1^n \frac{1}{n}\delta_{x_i} \in A\}) > 0$,*
then

$$\frac{1}{n}\ln\mu^n\left(\left\{x : \sum_1^n \frac{1}{n}\delta_{x_i} \in A\right\}\right) \geq -\frac{\mathrm{Ent}(\nu|\mu)}{\nu^n\left(\left\{x : \sum_1^n \frac{1}{n}\delta_{x_i} \in A\right\}\right)}$$

$$+\frac{1}{n}\ln\nu^n\left(\left\{x : \sum_1^n \frac{1}{n}\delta_{x_i} \in A\right\}\right) - \frac{1}{ne\nu^n\left(\left\{x : \sum_1^n \frac{1}{n}\delta_{x_i} \in A\right\}\right)}.$$

Proof. Set $\nu = f\mu$, $\nu^n = h\mu^n$ and

$$B = \left\{x : h(x) > 0, \ \sum_1^n \frac{1}{n}\delta_{x_i} \in A\right\}.$$

Notice that $\nu^n(\{h = 0\}) = \int_{\{h=0\}} h\,d\mu^n = 0$, so that still $\nu^n(B) > 0$. Denoting by $\nu_B^n(C) = \nu^n(B \cap C)/\nu^n(B)$ the normalized restriction of ν^n to B, using Jensen's inequality with the concave function $z \mapsto \ln z$ we can estimate

$$\ln\mu^n\left(\left\{x : \sum_1^n \frac{1}{n}\delta_{x_i} \in A\right\}\right) \geq \ln\mu^n(B) = \ln\left[\nu^n(B)\int h^{-1}\,d\nu_B^n\right]$$

$$\geq \ln\nu^n(B) + \int \ln h^{-1}\,d\nu_B^n$$

$$\geq \ln\nu^n(B) - \frac{\mathrm{Ent}(\nu^n|\mu^n)}{\nu^n(B)}$$

$$+ \frac{1}{\nu^n(B)}\int_{(\mathbb{R}^d)^n\setminus B} h\ln h\,d\mu^n.$$

But, since $h(x_1, \ldots, x_n) = f(x_1)\cdots f(x_n)$, we have

$$\mathrm{Ent}(\nu^n|\mu^n) = \int \ln h\,d\nu^n = n\int \ln f\,d\nu = n\mathrm{Ent}(\nu|\mu)$$

and we get

$$\ln\mu^n\left(\left\{x : \sum_1^n \frac{1}{n}\delta_{x_i} \in A\right\}\right) \geq \ln\nu^n(B) - \frac{n\mathrm{Ent}(\nu|\mu)}{\nu^n(B)}$$

$$+ \frac{1}{\nu^n(B)}\int_{(\mathbb{R}^d)^n\setminus B} h\ln h\,d\mu^n.$$

Now, remember that $x \ln x \geq -1/e$ to obtain

$$\ln \mu^n \left(\left\{ x : \sum_1^n \frac{1}{n} \delta_{x_i} \in A \right\} \right) \geq \ln \nu^n(B) - \frac{n \mathrm{Ent}(\nu|\mu)}{\nu^n(B)} - \frac{1}{e\nu^n(B)}.$$

Finally, $\nu^n(B) = \nu^n \left(\{ x : \sum_1^n \frac{1}{n} \delta_{x_i} \in A \} \right)$ and the lemma is proved. □

We have now all tools to prove (2.14) and Gozlan's theorem. From the previous lemma with $A = \{ \nu : W_2(\nu, \mu) > u \}$ we get the lower bound

$$\liminf_{n \to \infty} \frac{1}{n} \mu^n \left(\left\{ x : W_2 \left(\sum_1^n \frac{1}{n} \delta_{x_i}, \mu \right) > u \right\} \right) \geq -\mathrm{Ent}(\nu|\mu) \quad (2.15)$$

for any $\nu \ll \mu$ with $W_2(\nu, \mu) > u$, taking into account that the law of large numbers for empirical measures (statement (c) above) gives

$$\nu^n \left(\left\{ x : W_2 \left(\sum_1^n \frac{1}{n} \delta_{x_i}, \nu \right) > W_2(\mu, \nu) - u \right\} \right) \to 0$$

hence

$$\nu^n \left(\left\{ x : W_2 \left(\sum_1^n \frac{1}{n} \delta_{x_i}, \mu \right) > u \right\} \right) \to 1.$$

This proves (2.14). Now, let m_n be medians of $W_2(\sum_1^n \frac{1}{n} \delta_{x_i}, \mu)$ under μ^n and recall that, by the same law of large numbers, $m_n \to 0$. If $u > 0$ and $\sqrt{n}(u - m_n) \geq r_0$ (and this is true for sufficiently large n) we can use (2.11) with $r = \sqrt{n}(u - m_n)$ and the $1/\sqrt{n}$-Lipschitz continuity of $x \mapsto W_2(\sum_1^n 1/n \delta_{x_i}, \mu)$ to get

$$\mu^n \left(\left\{ x : W_2 \left(\sum_1^n \frac{1}{n} \delta_{x_i}, \mu \right) > u \right\} \right)$$

$$\leq \mu^n \left(\left(\left\{ x : \sqrt{n} W_2 \left(\sum_1^n \frac{1}{n} \delta_{x_i}, \mu \right) > \sqrt{n} m_n \right\} \right)_{\sqrt{n}(u - m_n)} \right)$$

$$\leq b e^{-a(\sqrt{n}(u - m_n) - r_0)^2}.$$

Taking logarithms, this provides the $\limsup$ inequality

$$\limsup_{n \to \infty} \frac{1}{n} \mu^n \left(\left\{ x : W_2 \left(\sum_1^n \frac{1}{n} \delta_{x_i}, \mu \right) > u \right\} \right) \leq -au^2. \quad (2.16)$$

Combining (2.15) and (2.16) gives

$$au^2 \leq \mathrm{Ent}(\nu|\mu) \qquad \text{for all } \nu \ll \mu \text{ with } W_2(\nu, \mu) > u$$

which is an equivalent formulation of $T_2(1/a)$. □

2.5 T_2 and LS in any dimension

Up to now, we have proved that the LS inequalities implies dimension-free Gaussian concentration (tensorisation and Herbst argument) and also that the quadratic transportation cost inequality T_2 is equivalent to dimension-free Gaussian concentration (arguments of Marton, Talagrand and Gozlan's theorem). Therefore we have completed the first objective of this section on concentration, namely to show

$$LS \Longrightarrow T_2 \Longleftrightarrow \text{dimension-free Gaussian concentration.}$$

Our remaining task is to show that $T_2 \not\Longrightarrow LS$, which somehow requires to understand what has been lost in deriving $LS \Longrightarrow T_2$. Actually, tracing the constants shows that we have proved

$$\mu \in LS(C) \Longrightarrow \mu \in T_2(2C)$$

by using Gozlan's theorem. The above implication was first discovered by Otto and Villani [37], before Gozlan's result, and with a more direct argument. For further use, we shall however present another proof due to Bobkov, Gentil and Ledoux [6]. We start with recalling an equivalent (dual) formulation of the $T_2(a)$ property based on inf-convolutions.

Proposition 2.12 (Bobkov and Götze [7]). *For $a > 0$, $\mu \in T_2(a)$ if and only if*

$$\int e^{(Qg)/a} \, d\mu \le e^{\int g \, d\mu/a} \qquad \text{for all } g \text{ bounded Lipschitz} \qquad (2.17)$$

where $Qg(x) = \inf_y \left\{ g(y) + |x - y|^2 \right\}$.

In order to prove it, let us rewrite the T_2 inequality, namely $T_2(f\mu,\mu) \le a\text{Ent}_\mu(f)$ in the dual way (see (4.1) and (2.10)):

$$T_2(f\mu, \mu) = \sup \left\{ \int h f \, d\mu - \int g \, d\mu : h(x) - g(y) \le |x - y|^2 \right\}$$

$$= \sup \left\{ \int Qgf \, d\mu - \int g \, d\mu \right\}$$

$$= \sup \left\{ \int Qgf \, d\mu : \int g \, d\mu = 0 \right\}$$

because Qg is the largest function less than $g(y) + | \cdot - y|^2$ for all y and $Q(g + k) = Qg + k$ for all $k \in \mathbb{R}$. It follows that $\mu \in T_2(a)$ if and only if, for all probability densities f,

$$\sup \left\{ \int Qgf \, d\mu : \int g d\mu = 0 \right\} \le \sup \left\{ \int fh d\mu : \int e^{h/a} d\mu \le 1 \right\}. \quad (2.18)$$

By Hölder's inequality $\{h; \int \exp(h/a) \, d\mu \leq 1\}$ is convex, so a heuristic separation argument suggests that the above inequality should be equivalent to

$$\int g \, d\mu = 0 \implies \int e^{(Qg)/a} \, d\mu \leq 1. \tag{2.19}$$

A rigorous argument is as follows: if (2.19) is valid, then (2.18) is true simply because one may take $h = Qg$. Conversely, if $\int g \, d\mu = 0$ and (2.18) holds, the inequality that we obtain by choosing

$$f = \frac{e^{\frac{Qg}{a}}}{\int e^{\frac{Qg}{a}} \, d\mu} \quad \text{and} \quad h = a \log f$$

boils down to $\int e^{\frac{Qg}{a}} \, d\mu \leq 1$. Eventually, adding constants to g shows that (2.19) is equivalent to (2.17). □

We can now prove the implication

$$LS(C) \implies T_2(2C).$$

Taking the previous proposition into account, we assume $LS(C)$ and we need to prove (2.17) with $a = 2C$. By an approximation argument we shall prove this implication when $g \in C_b$ and we follow an argument similar to the one used by Herbst, defining

$$H(\lambda) := \int e^{Q(\lambda g)/2C} \, d\mu \qquad \forall \lambda \geq 0. \tag{2.20}$$

Notice that, by dominated convergence, $H \in C([0, +\infty))$ and $H(0) = 1$.

Let us recall a few facts from the theory of Hamilton-Jacobi equations: if we set

$$q_t g(x) := \inf_y g(y) + \frac{|x - y|^2}{2t},$$

then $q_t g \in C([0, +\infty) \times \mathbb{R}^d)$, is Lipschitz in all sets $(\delta, +\infty) \times \mathbb{R}^d$ for $\delta > 0$, and solves (see for instance [17])

$$\begin{cases} \partial_t (q_t g) + \dfrac{|\nabla (q_t g)|^2}{2} = 0 \\ q_0 g(x) = g(x) \end{cases} \tag{2.21}$$

$\mathscr{L}^{d+1}$-a.e. in $(0, +\infty) \times \mathbb{R}^d$. Also, the operators Q and q are related for $\lambda > 0$ by

$$Q(\lambda g) = \inf_y \lambda \left(g(y) + \frac{|x - y|^2}{2\lambda/2} \right) = \lambda q_{\lambda/2} g.$$

From this identity and the continuity of q_t at the origin we get

$$\frac{d}{d\lambda}Q(\lambda g)\Big|_{\lambda=0} = g,$$

so that dominated convergence gives

$$\frac{d}{d\lambda}H(\lambda)\Big|_{\lambda=0} = \frac{\int g\,d\mu}{2C}. \tag{2.22}$$

In addition, (2.21) gives

$$\frac{d}{d\lambda}Q(\lambda g) = q_{\lambda/2}g - \frac{\lambda}{4}|\nabla q_{\lambda/2}g|^2$$

for $\mathscr{L}^{d+1}$-a.e. (t, x). It turns out that, for $\mathscr{L}^1$-a.e. $\lambda > 0$, we have the differential relation

$$\frac{d}{d\lambda}\lambda Q(\lambda g)(x) = Q(\lambda g)(x) - \frac{1}{4}|\nabla Q(\lambda g)(x)|^2 \quad \text{for } \mathscr{L}^d\text{-a.e. } x. \tag{2.23}$$

Now, differentiating for $\lambda > 0$ in (2.20) (differentiation under the integral sign is not a problem, by the local in time uniform Lipschitz property) we have

$$\lambda H'(\lambda) = \int \frac{1}{2C}e^{(Q(\lambda g))/2C}\left(Q(\lambda g) - \frac{1}{4}|\nabla Q(\lambda g)|^2\right)d\mu$$

for $\mathscr{L}^1$-a.e. λ and we can apply $LS(C)$ with $f^2 = e^{Q(\lambda g)/2C}$ to control the first term in the right hand side, to get

$$\lambda H'(\lambda) \le H(\lambda)\ln H(\lambda) + 2C\int|\nabla f|^2\,d\mu - \frac{1}{8C}\int e^{Q(\lambda g)/2C}|\nabla Q(\lambda g)|^2\,d\mu.$$

Now, the remarkable fact is that the two last terms cancel and we conclude that $\lambda H'(\lambda) \le H(\lambda)\ln H(\lambda)$, i.e.

$$\frac{d}{d\lambda}\left(\frac{\ln H(\lambda)}{\lambda}\right) \le 0.$$

Recall that this inequality holds for $\mathscr{L}^1$-a.e. $\lambda > 0$; but since H is locally Lipschitz in $(0, +\infty)$ we can integrate the differential inequality, taking (2.22) into account, to get

$$\frac{\ln H(\lambda)}{\lambda} \le \lim_{\lambda\downarrow 0}\frac{\ln H(\lambda)}{\lambda} = H'(0) = \frac{\int g\,d\mu}{2C}.$$

Setting $\lambda = 1$, this proves that $\int e^{(Qg)/2C}\,d\mu \le e^{\int g\,d\mu/2C}$.   □

As noted by Cattiaux and Guillin, the above proof uses the log-Sobolev inequality for rather specific functions of the form $f^2 = e^{Q(\lambda g)/2C}$. This allows the following refinement:

Theorem 2.13 (Cattiaux-Guillin [12]). *Let* $\mu \in \mathscr{P}(\mathbb{R}^d)$. *Assume that there exists* $\eta, C_\eta > 0$ *and* $x_0 \in \mathbb{R}^d$ *such that every function* f *satisfying for all* x

$$f^2(x) \le \left(\int f^2 \, d\mu \right) e^{2\eta \left(|x-x_0|^2 + \int |y-x_0|^2 d\mu(y) \right)},$$

also verifies

$$\mathrm{Ent}_\mu(f^2) \le 2C_\eta \int |\nabla f|^2 \, d\mu.$$

Then $\mu \in T_2(\max\{2C_\eta, 1/\eta\})$.

Proof. Set $\tilde{\eta} = \min(\eta, 1/2C_\eta)$. Let g be a function with $\int g \, d\mu = 0$. Define

$$H(\lambda) = \int e^{\tilde{\eta} Q(\lambda g)} d\mu.$$

In view of Proposition 2.12, our task is to prove that $H(1) \le 1$. Assume on the contrary that $H(1) > 1 = H(0)$. Let λ_0 be the smallest (nonnegative) number such that $H(\lambda_0) = 1$ and $H(\lambda) \ge 1$ for all $\lambda \in [\lambda_0, 1]$. The main observation is that

$$Q(\lambda g)(x) = \inf_y \{\lambda g(y) + |x - y|^2\} \le \int (\lambda g(y) + |x - y|^2) \, d\mu(y)$$

$$= \int |x - x_0 + x_0 - y|^2 d\mu(y) \le 2|x - x_0|^2 + 2 \int |x_0 - y|^2 d\mu(y).$$

So for $\lambda \in [\lambda_0, 1]$, using $H(\lambda) \ge 1$ and $\tilde{\eta} \le \eta$ yields

$$e^{\tilde{\eta} Q(\lambda g)(x)} \le e^{2\tilde{\eta} \left(|x-x_0|^2 + \int |x_0-y|^2 d\mu(y) \right)}$$

$$\le \left(\int e^{\tilde{\eta} Q(\lambda g)} d\mu \right) e^{2\eta \left(|x-x_0|^2 + \int |x_0-y|^2 d\mu(y) \right)}.$$

Hence by hypothesis, the nonnegative function f such that $f^2 = e^{\tilde{\eta} Q(\lambda g)}$ satisfies

$$\mathrm{Ent}_\mu(f^2) \le 2C_\eta \int |\nabla f|^2 \, d\mu.$$

Following the argument of Bobkov-Gentil-Ledoux (and using $\tilde{\eta} \le 1/(2C_\eta)$) we get that $\frac{d}{d\lambda}(\frac{1}{\lambda} \log H(\lambda)) \le 0$ for all $\lambda \in [\lambda_0, 1]$. Consequently $\log H(1) \le \frac{1}{\lambda_0} \log H(\lambda_0) = 0$. □

2.6   More on T_2 and LS in one dimension

The latter statement suggests that the T_2 inequality is substantially weaker than the log-Sobolev inequality. It is actually the starting point of the first counterexample to the implication $T_2 \implies LS$, provided by Cattiaux and Guillin. In this section, we give a proof of the following statement.

Theorem 2.14 (Cattiaux-Guillin [12]). *For $\beta \in (2, 5/2]$ the probability measure μ_β defined on $\mathbb{R}$ by $\mu_\beta = \frac{1}{z_\beta} e^{-V_\beta(|x|)} \mathscr{L}^1$, where*

$$V_\beta(x) = x^3 + 3x^2 \sin^2 x + x^\beta \tag{2.24}$$

and $z_\beta > 0$ is a normalization constant, satisfies a quadratic transportation cost inequality $T_2(C_\beta)$ some constant C_β but does not verify the log-Sobolev inequality (for any finite constant).

The proof will use some very precise criteria for measures on $\mathbb{R}$ to satisfy Sobolev type inequalities. In the case of the log-Sobolev inequalities, Bobkov and Götze [7] have provided a way to compute, up to universal constants the (possibly infinite) best number C such that a probability measure on $\mathbb{R}$ belongs to $LS(C)$. We are going to present a simplified approach, developed with Roberto in [5]. It provides better numerical constants and allows extensions

Theorem 2.15 ([7], [5]). *Let $\mu \in \mathscr{P}(\mathbb{R})$, let $n \in L^1(\mathbb{R})$ nonnegative and let $\nu = n\mathscr{L}^1$. Then, the best constant $C \in [0, +\infty]$ such that for all f*

$$\mathrm{Ent}_\mu(f^2) \le C \int |f'|^2 \, d\nu \tag{2.25}$$

satisfies $\max\{b_+, b_-\} \le C \le 16 \max\{b_+, b_-\}$, where

$$b_+ = \sup_{x > m} \mu\left([x, \infty)\right) \ln\left(1 + \frac{1}{2\mu([x, \infty))}\right) \int_m^x \frac{1}{n(t)} \, dt$$

$$b_- = \sup_{x < m} \mu\left((-\infty, x]\right) \ln\left(1 + \frac{1}{2\mu((-\infty, x])}\right) \int_x^m \frac{1}{n(t)} \, dt$$

and m is any median of μ.

Corollary 2.16. *Let $\mu = e^{-V} \mathscr{L}^1$, with V continuous and satisfying*

(i) $\liminf_{|x| \to \infty} \mathrm{sign}(x) V'(x) > 0$;

(ii) $\limsup_{|x| \to \infty} \frac{|V''(x)|}{(V'(x))^2} < 1$.

Then

$$\exists C \text{ such that } \mu \text{ satisfies } LS(C) \quad \Longleftrightarrow \quad \limsup_{|x| \to \infty} \frac{V(x)}{(V'(x))^2} < \infty.$$

The corollary is a direct consequence of Theorem 2.15: indeed, (i) and (ii) imply

$$\int_x^\infty e^{-V(y)} dy \sim \frac{e^{-V(x)}}{V'(x)} \tag{2.26}$$

because the ratio of their derivatives goes like $(1 + V''/|V'|^2)$; analogously

$$\int_m^x e^V(y) dy \sim \frac{e^{V(x)}}{V'(x)} \tag{2.27}$$

because the ratio of their derivatives goes like $(1 - V''/|V'|^2)$; from (2.26) and (2.27) we infer that the finiteness of b_+ and b_- is equivalent to

$$\limsup_{\pm\infty} \frac{V - \ln V'}{|V'|^2} < \infty$$

which, in turn, is equivalent to $\limsup V/|V'|^2 < \infty$. $\qquad \square$

As a consequence of the corollary, the measure μ corresponding to $V(x) = |x|^\alpha$ satisfies a LS inequality if and only if $\alpha \geq 2$. It is also useful to introduce another relevant property of measures μ, the well-known *Poincaré* or spectral gap inequality.

Definition 2.17 (Poincaré inequality). If $\mu \in \mathscr{P}(\mathbb{R}^d)$ with $\mu \ll \mathscr{L}^d$, we say that μ satisfies a Poincaré inequality with constant C, and write $\mu \in PI(C)$, if

$$\text{Var}_\mu(f) \leq C \int |\nabla f|^2 d\mu$$

for any locally Lipschitz function $f : \mathbb{R}^d \to \mathbb{R}$.

It is known that a Poincaré inequality implies dimension-free exponential concentration (see *e.g.* [26]) which is less stringent than Gaussian concentration. Also Otto and Villani [37] have shown that $\mu \in T_2(2C) \implies \mu \in PI(C)$. A workable criterion is also available for a probability measure on the real line to satisfy a Poincaré inequality. It is actually simpler than for LS. The following result is quoted in [2]

Theorem 2.18. *Let $\mu \in \mathscr{P}(\mathbb{R})$. The best constant C with $\mu \in PI(C)$ satisfies*

$$\max\{B_+, B_-\}/2 \leq C \leq 4\max\{B_+, B_-\},$$

where

$$B_+ = \sup_{x>m} \mu\left([x, \infty)\right) \int_m^x \frac{1}{n(t)}\, dt$$

$$B_- = \sup_{x<m} \mu\left((-\infty, x])\right) \int_x^m \frac{1}{n(t)}\, dt$$

and m is any median of μ.

The main ingredient in the proof of Theorems 2.15 and 2.18 is a Hardy inequality in weighted spaces, stated next.

Theorem 2.19. *Let μ, $\nu = n\mathcal{L}^1$ be finite measures in $[m, +\infty)$. Then, the best constant C such that*

$$\int_m^\infty f^2\, d\mu \le C \int_m^\infty |f'|^2\, d\nu$$

for all f with $f(m) = 0$ satisfies $B \le C \le 4B$, where

$$B = \sup_{x>m} \mu\left([x, \infty)\right) \int_m^x \frac{1}{n(s)}\, ds. \tag{2.28}$$

For measures with densities the above statement is due to Talenti [45], Tomaselli [46] and Artola (unpublished). Muckenhoupt [36] has shown that it is valid of arbitrary finite measures, provided one defines the quantity n in (2.28) as the density of the absolutely continuous part of ν. We will not need such generality.

Proof. The lower bound in the latter theorem is not hard: if we consider $x > m$ and a function f with $f \equiv 1$ in $[x, \infty)$ and $f(m) = 0$, we have

$$\int_m^\infty f^2\, d\mu \ge \mu\left([x, +\infty)\right),$$

hence C is bounded below by

$$\mu\left([x, +\infty)\right)\left(\int_m^x |f'(t)|^2 n(t)\, dt\right)^{-1}.$$

We are thus led to the problem of minimizing $\int_m^x |f'(t)|^2 n(t)\, dt$ with the constraints $f(m) = 0$, $f(x) = 1$. By the Cauchy-Schwartz inequality

$$1 \le \left|\int_m^x f'(t)\, dt\right|^2 \le \int_m^x |f'(t)|^2 n(t)\, dt \int_m^x \frac{1}{n(t)}\, dt$$

hence the minimum can be bounded from below by $\left(\int_m^x \frac{1}{n(t)}\,dt\right)^{-1}$; this lower bound is sharp, considering

$$f(y) = \frac{\int_m^y \frac{1}{n(t)}\,dt}{\int_m^x \frac{1}{n(t)}\,dt} \qquad y \in [m, x],$$

and we conclude that a lower bound for C is given by the constant B in (2.28). Let us emphasize for further use that we have shown

$$\inf\left\{\int_m^x |f'(t)|^2 n(t)\,dt; \; f(m) = 0 \text{ and } f(x) = 1\right\}$$

$$= \left(\int_m^x \frac{1}{n(t)}\,dt\right)^{-1}. \tag{2.29}$$

The proof of the upper bound in Theorem 2.19, with the sharp constant 4, is rather tricky. It can be found in the above references or in [2]. We give here a simple and natural argument which gives a worse numerical constant. First, it is easy to see that it suffices to prove the Hardy inequality for non-decreasing functions f with $f(m) = 0$. Next we proceed to a dyadic decomposition into level sets, writing

$$\int f^2 d\mu = \int_0^{+\infty} \mu(\{f^2 \geq t\})\,dt \leq \sum_{k\in\mathbb{Z}} \int_{2^k}^{2^{k+1}} \mu(\{f^2 \geq t\})\,dt$$

$$\leq \sum_{k\in\mathbb{Z}} 2^k \mu(\{f^2 \geq 2^k\}).$$

For a non-decreasing (continuous) function f with $f(m) = 0$, the level-set $\{f^2 \geq 2^k\}$ is either empty or of the form $[a_k, +\infty[$ for some $a_k > m$. Combining the definition (2.28) and (2.29) gives that for any locally Lipschitz function g on $[m, a_k]$ with $g(m) = 0$ and $g(a_k) = 1$ it holds

$$\mu([a_k, +\infty[) \leq B \int_m^{a_k} (g'(t))^2 n(t)\,dt.$$

If we choose

$$g = \left(\frac{f - 2^{\frac{k-1}{2}}}{2^{\frac{k}{2}} - 2^{\frac{k-1}{2}}}\right)_+,$$

which vanishes before a_{k-1} and is equal to 1 at a_k, we obtain for non-empty level-set

$$\mu(\{f^2 \geq 2^k\}) = \mu([a_k, +\infty[) \leq B \int_{a_{k-1}}^{a_k} \frac{f'(t)^2}{(2^{\frac{k}{2}} - 2^{\frac{k-1}{2}})^2} n(t)\,dt.$$

Summing over non-empty level sets and using the above upper bound on $\int f^2 d\mu$ readily yields

$$\int f^2 d\mu \leq \frac{B}{\left(1 - \dfrac{1}{\sqrt{2}}\right)^2} \int (f')^2 n.$$

$\square$

It is also worthwhile mentioning that Theorem 2.19 admits a useful immediate extension, put forward by Bobkov and Götze [7]: if $\mathcal{G}$ is a family of nonnegative functions and we define

$$\Phi(f) := \sup_{g \in \mathcal{G}} \int fg \, d\mu$$

then (under the same assumptions on μ and ν) the best constant C such that

$$\Phi(f^2) \, d\mu \leq C \int_m^\infty |f'|^2 \, d\nu$$

is true for all f with $f_{|(-\infty,m]} = 0$ satisfies $B \leq C \leq 4B$, where now B is defined by

$$B = \sup_{x > m} \Phi(\chi_{[x,\infty)}) \int_m^x \frac{1}{n(s)} \, ds.$$

This formulation seems quite adapted for log-Sobolev inequalities thanks to the dual formula for the entropy (4.1)

$$\mathrm{Ent}_\mu(f) = \sup \left\{ \int fg \, d\mu : g \in C_b, \int e^g \, d\mu \leq 1 \right\}.$$

However the only non-negative test function g in this formula is the zero function. As shown in [5], some more work allows to turn this failed attempt into a correct proof. We explain this next, focusing on the lower bound on the best log-Sobolev constant (this is all we need to prove Theorem 2.14).

Proof of the lower bound in Theorem 2.15. Let us assume that (2.25) holds for μ and let m be a median of μ. We consider a smooth function f identically equal to 0 on $(-\infty, m]$ and notice that, by the entropy duality formula,

$$C \int_m^\infty |f'|^2 \, d\nu \geq \sup \left\{ \int_m^\infty f^2 g \, d\mu : \int e^g \, d\mu \leq 1 \right\}.$$

Considering functions g identically equal to $-\infty$ on $(-\infty, m)$, we reduce ourselves to the space $X = [m, +\infty)$ and bound from below the above supremum by

$$\sup \left\{ \int_m^\infty f^2 g \, d\mu \, : \, g \geq 0, \, \int_m^\infty e^g \, d\mu \leq 1 \right\}.$$

Now we can apply Theorem 2.19 to all measures $g\mu$, to obtain

$$C \geq \sup \left\{ \sup_{x > m} \int_m^\infty \chi_{[x,+\infty)} g \, d\mu \int_m^x \frac{1}{n(s)} \, ds \, : \, g \geq 0, \, \int_m^\infty e^g \, d\mu \leq 1 \right\}.$$

Commuting the sup and using Lemma 2.20 below with $A = [x, \infty)$ and $\sigma = \mu$ we conclude that $C \geq B_+$. The proof of the lower bound $C \geq B_-$ is analogous. □

Lemma 2.20. *If σ is a measure in X with $\sigma(X) \leq 1/2$, then for $A \in \mathcal{B}(X)$ with $\sigma(A) > 0$ we have*

$$\sup \left\{ \int_A h \, d\sigma \, : \, h \geq 0, \, \int e^h \, d\sigma \leq 1 \right\} = \sigma(A) \ln \left(1 + \frac{1}{2\sigma(A)} \right).$$

The proof of the lemma is elementary: first, we see that the maximization problem is equivalent to

$$\sup \left\{ \int_A h \, d\sigma \, : \, h \geq 0, \, \int_A e^h \, d\sigma \leq 1 - \sigma(A^c) \right\}.$$

Now, working in the measure space A, we see that the constant function h with $e^h = (1 - \sigma(A^c))/\sigma(A)$ is admissible (by our assumption that $\sigma(A) \leq 1/2$) and therefore the supremum can be bounded from below by

$$\sigma(A) \ln \frac{1 - \sigma(A^c)}{\sigma(A)} \geq \sigma(A) \ln \frac{\frac{1}{2} + \sigma(A)}{\sigma(A)}.$$

Using Jensen's inequality one can easily prove that the function h above is indeed a maximizer. □

Proof of Theorem 2.18. Let m be a median of μ. We start with the upper bound on the best Poincaré constant: thanks to Theorem 2.19

$$\mathrm{Var}_\mu(f) \leq \int (f - f(m))^2 \, d\mu = \int_{-\infty}^m (f - f(m))^2 \, d\mu + \int_m^\infty (f - f(m))^2 \, d\mu$$

$$\leq 4B_+ \int_m^\infty |f'|^2 \, d\mu + 4B_- \int_{-\infty}^m |f'|^2 \, d\mu.$$

Next we turn to the lower bound. Assume that $\mu \in PI(C)$. Let f such that $f_{|(-\infty,m]} = 0$, then

$$\left(\int f\, d\mu \right)^2 = \left(\int f\, \chi_{]m,+\infty)}\, d\mu \right)^2 \leq \frac{1}{2} \int f^2 d\mu.$$

Consequently

$$C \int (f')^2\, d\mu \geq \operatorname{Var}_\mu(f) = \int f^2 d\mu - \left(\int f\, d\mu \right)^2 \geq \frac{1}{2} \int f^2 d\mu.$$

Since this is true for all f vanishing before m, we conclude by Theorem 2.19 that $2C \geq B_+$. A similar argument yields $2C \geq B_-$. $\qquad\square$

With the previous tools in hands, we are ready to attack the problem of finding a good sufficient condition for a measure to satisfy the T_2 inequality. The strategy is to use Theorem 2.13 and to prove a log-Sobolev inequality restricted to the class of functions involved there. In the sufficient condition, we will include some necessary conditions for T_2 which are easy to check as Gaussian integrability or Poincaré inequality. We will use the following lemma, which we quote from [35]. It allows a more compact argument than in [12] but gives worse dependencies in the parameters.

Lemma 2.21. *Let $K > 1$ and $f \in L_2(\mu)$. Then*

$$\operatorname{Ent}_\mu(f^2) \leq (K+1)^2 \operatorname{Var}_\mu(f) + \int_{\{f^2 \geq K^2 \int f^2 d\mu\}} f^2 \ln\left(\frac{f^2}{\int f^2 d\mu} \right) d\mu.$$

Proof. Assume, without loss of generality that $\int f^2 d\mu = 1$ and $\int f\, d\mu \geq 0$. Let $\Phi_1(t) = t \ln t - t + 1$. Since $\int f^2 d\mu = 1$, $\operatorname{Ent}_\mu(f^2) = \int \Phi_1(f^2)\, d\mu$. Note that for $t > 1$, $\Phi_1(t) \leq t \ln t$, hence

$$\int_{\{f^2 > K^2\}} \Phi_1(f^2)\, d\mu \leq \int_{\{f^2 \geq K^2\}} f^2 \ln\left(f^2\right) d\mu.$$

Next, when $|f| \leq K$, using the inequality $\ln u \leq u - 1$, we obtain

$$\Phi_1(f^2) \leq (f^2 - 1)^2 \leq (K+1)^2 (f-1)^2.$$

Hence $\int_{\{f^2 \leq K^2\}} \Phi_1(f^2)\, d\mu \leq (K+1)^2 \int (f-1)^2 d\mu$. Recall that $\int f^2 d\mu = 1$, and $\int f\, d\mu \geq$ so that expanding the square and using Cauchy-Schwarz, we get

$$\int (f-1)^2 d\mu = 2 \int f^2 d\mu - 2 \left(\int f\, d\mu \right) \sqrt{\int f^2 d\mu}$$

$$\leq 2 \left(\int f^2 d\mu - \left(\int f\, d\mu \right)^2 \right).$$

Combining the above estimates readily gives the claim. □

Assume for simplicity that 0 is a median of μ. Let us try and prove a log-Sobolev inequality for μ restricted to functions satisfying

$$f(x)^2 \leq \left(\int f^2 d\mu \right) e^{2\eta(1+x^2)}, \quad \forall x \in \mathbb{R}.$$

We will also assume that μ satisfies a Poincaré inequality with constant denoted C_P. Without loss of generality, we suppose $\int f^2 d\mu = 1$. Let $K > 2e^\eta$. The previous lemma, the spectral gap inequality and the hypothesis on f give

$$\mathrm{Ent}_\mu(f^2) \leq (K+1)^2 \mathrm{Var}_\mu(f) + \int_{\{f^2 \geq K^2\}} f^2 \ln\left(f^2\right) d\mu$$

$$\leq (K+1)^2 C_P \int (f')^2 d\mu + \eta \int_{\{f^2 \geq K^2\}} f(x)^2 (1+x^2) d\mu(x).$$

It remains to upper bound the latter term by a multiple of $\int (f')^2 d\mu$. Consider the function $\psi : \mathbb{R} \to \mathbb{R}^+$ defined by

$$\psi(t) = \begin{cases} 0 & \text{if } |t| \leq K/2 \\ 2|t| - K & \text{if } |t| \in [K/2, K] \\ |t| & \text{if } |t| \geq K. \end{cases}$$

By hypothesis $|f(0)| \leq e^\eta < K/2$ so that $\psi(f(0)) = 0$. Hence we may apply Theorem 2.19 with measures $(1+x^2)d\mu(x)$ and μ, on $\mathbb{R}^+$ and on $\mathbb{R}^-$ to the function $\psi \circ f$. Setting (n being the density of μ)

$$A_+ = \sup_{x>0} \int_x^\infty (1+t^2)\, d\mu(t) \int_0^x \frac{1}{n(t)}\, dt$$

$$A_- = \sup_{x<0} \int_{-\infty}^x (1+t^2)\, d\mu(t) \int_x^0 \frac{1}{n(t)}\, dt,$$

we obtain the following estimate

$$\int_{\{f^2 \geq K^2\}} f(x)^2(1+x^2)d\mu(x) \leq \int (\psi(f(x))^2(1+x^2)d\mu(x)$$

$$\leq 4\max\{A_+, A_-\} \int ((\psi \circ f)')^2 d\mu \leq 16\max\{A_+, A_-\} \int (f')^2 d\mu,$$

where in the last inequality we have used that $|\psi'| \leq 2$ a.e.

Summing up, we have proved the following result from [12]:

Theorem 2.22. *Let $\mu = n\mathcal{L}^1 \in \mathcal{P}(\mathbb{R})$, with median m, be satisfying*

$$\sup_{x>m} \int_x^\infty (1+t^2)\,d\mu(t) \int_m^x \frac{1}{n(t)}\,dt < \infty$$

$$\sup_{x<m} \int_x^m (1+t^2)\,d\mu(t) \int_{-\infty}^x \frac{1}{n(t)}\,dt < \infty.$$

Then there exists C such that $\mu \in T_2(C)$.

Indeed, the finiteness of the two suprema above implies both the validity of a Poincaré inequality (by Theorem 2.18) and of the restricted LS inequality, sufficient to get the T_2 property thanks to Theorem 2.13. Arguing as in the proof of Corollary 2.16 we have also the following result:

Corollary 2.23. *Let $\mu = e^{-V}\mathcal{L}^1$, with V continuous, satisfying]*

  (i) $\liminf_{|x|\to\infty} \mathrm{sign}(x)V'(x) > 0$;
  (ii) $\limsup_{|x|\to\infty} \frac{|V''(x)|}{(V'(x))^2} < 1$;
  (iii) $\limsup_{|x|\to\infty} \frac{x^2}{|V'(x)|^2} < \infty$.

Then there exists $a > 0$ such that μ satisfies $T_2(a)$.

The discrepancy between the criterion for LS inequality (Corollary 2.16) and the above necessary condition for T_2 allows to find a measure satisfying T_2 and not LS. The proof of Theorem 2.14 is now just a matter of elementary calculations. Let us however make a few comments which have lead to the choice of a counterexample as given in (2.24). Firstly, the potential V should not be of quadratic order otherwise condition (iii) above implies the asymptotic boundedness of $V/(V')^2$ and thus a LS inequality. Secondly, a result of Wang (stated in Theorem 2.24 below) suggests to look for a counterexample with $\liminf V'' = -\infty$. Indeed, it asserts that if there exists $\epsilon, k > 0$ such that $V'' \geq -k$ pointwise and $\int e^{(k/2+\epsilon)|x|^2} d\mu(x) < \infty$, then μ satisfies a LS inequality. So if V''

is bounded from below and if $\mu = e^{-V}\mathscr{L}^1$ satisfies $T_2(C)$ for a small enough C then by concentration for the Lipschitz function $|x|$ the above integrability condition and thus LS hold. See also [37] for $T_2 \implies LS$ when the Hessian of the potential is bounded from below, with more precise conditions on the constants.

Finally, let us mention the more recent papers [G, G3] where other criteria are developed for general transportation cost inequalities.

2.7 Consequences of curvature

To conclude this section on concentration, we present a useful criterion for LS inequalities in any dimension. It allows in particular to prove LS and concentration for the Gaussian measure, with the optimal constants (which we could not reach with the previous one-dimensional criteria) and thus to make these notes more self-contained.

Theorem 2.24. *Let (M, ρ) be a Riemannian manifold endowed with the probability measure $\mu(dx) = e^{-V(x)}\mathrm{vol}(dx)$, and fix $\lambda \in \mathbb{R}$. Assume that $\mathrm{Ric}_x + \mathrm{Hess}_x\, V \geq \lambda\,\mathrm{Id}$ for all $x \in M$. Then:*

(a) *if $\lambda > 0$, then $\mu \in LS(1/\lambda)$ (Bakry-Emery [3]);*
(b) *if $\lambda = -K < 0$ and there exist $x_0 \in M$ and $\varepsilon > 0$ such that $\int e^{\frac{K+\varepsilon}{2}d(x,x_0)^2}\, d\mu(x) < +\infty$ where d denotes the geodesic distance, then there exists $C > 0$ such that $\mu \in LS(C)$ (Wang [49]).*

The original proofs of the statement involve interpolation by the evolution semigroup of generator $L = \Delta - \nabla v \cdot \nabla$. We will rather present a transportation proof. It is taken from Cordero's paper [13] for $\lambda > 0$ (which was inspired by the seminal work [37] of Otto and Villani) and [4] for $\lambda \leq 0$. The starting point, is Theorem 2.25 below, which can be viewed as an "above tangent" formulation of McCann's displacement convexity [29]. It was first put forward by Otto and Villani [37] and further developed by several authors, see *e.g.* [14, 47]. Before stating it, let us recall some important structural facts about optimal transportation: when μ, ν are Borel probability measures on $\mathbb{R}^d$ and $\nu \ll \mathscr{L}^d$, say compactly supported, the optimal π for the minimization problem

$$\inf_{\pi \in \Gamma(\mu,\nu)} \int_{\mathbb{R}^d \times \mathbb{R}^d} |x - y|^2 d\pi(x, y),$$

is actually supported by the graph of a map $T : \mathbb{R}^d \to \mathbb{R}^d$: it is of the form $d\pi(x, y) = \delta_{T(x)}(dy)\mu(dx)$ and marginal condition means that the image measure $T_\sharp\mu$ of μ by T is ν (meaning $\nu(A) = \mu(T^{-1}(A))$ for all $A \in \mathscr{B}(\mathbb{R}^d)$). By the Brenier theorem [10], T is the gradient of a convex function. This was extended by McCann [30,31] in particular to Riemannian manifolds, where one minimises the integral of the square geodesic distance and the optimal map is of the form $T(x) = \exp_x(\nabla\theta(x))$ with θ enjoying specific convexity-type properties.

Theorem 2.25. *Assume as above* $\text{Ric}_x + \text{Hess}_x\, V \geq \lambda\,\text{Id}$. *Given* $g\,d\mu$ *and* $h\,d\mu$ *two compactly supported probability measures on* M, *with* g *of class* C^1. *Let* T *denote the optimal transport map from* $g\,d\mu$ *to* $h\,d\mu$ *for the cost function* $d(x, y)^2$, $T(x) = \exp_x(\nabla\theta(x))$. *Then*

$$\text{Ent}_\mu(h) \geq \text{Ent}_\mu(g) + \int \langle\nabla\theta, \nabla g\rangle\, d\mu + \frac{\lambda}{2} W_2^2(h\,d\mu, g\,d\mu). \quad (2.30)$$

We just sketch a proof of Otto-Villani's theorem in the Euclidean case $M = \mathbb{R}^n$ and omit regularity issues. Since the image of $g\,d\mu$ by T is $h\,d\mu$, the change of variable formula gives

$$g(x)e^{-V(x)} = h(T(x))e^{-V(T(x))} \det DT(x),$$

so that taking the logarithm of both sides, and integrating with respect to $g\,d\mu$, one gets

$$\text{Ent}_\mu(g) = \text{Ent}_\mu(h) + \int [V(x) - V(T(x)) + \log(\det DT(x))]g(x)\, d\mu(x).$$

Then, since T is given by the gradient of a convex function, one has

$$\log(\det DT(x)) = \sum_i \log(\lambda_i(x)) \leq \sum_i (\lambda_i(x) - 1) = \text{tr}(DT(x) - I)$$

(here $\lambda_i(x) \geq 0$ denote the eigenvalues of $DT(x)$). Writing now $T(x)$ as $x + \nabla\theta(x)$ and exploiting the assumption $\text{Hess}_x\, V \geq \lambda Id$, we obtain

$$\text{Ent}_\mu(g) \leq \text{Ent}_\mu(h) + \int [V(x) - V(x - \nabla\theta(x)) + \Delta\theta(x)]g(x)\, d\mu(x)$$

$$\leq \text{Ent}_\mu(h) + \int \left(-\langle\nabla V, \nabla\theta\rangle - \frac{\lambda}{2}|\nabla\theta(x)|^2 - \Delta\theta(x)\right) g(x)\, d\mu(x).$$

To conclude it suffices to observe that

$$\int |\nabla\theta(x)|^2 g\, d\mu = \int |T(x) - x|^2 g\, d\mu = W_2^2(g\,d\mu, h\,d\mu),$$

and (recalling that $\mu = e^{-V}dx$)

$$\int [\langle\nabla V, \nabla\theta\rangle + \Delta\theta(x)]g\, d\mu = \int \text{div}(e^{-V}\nabla\theta)g\, dx$$

$$= -\int \langle\nabla\theta, \nabla g\rangle\, d\mu. \qquad \square$$

Remark 2.26. Choosing formally $g = 1$, Inequality (2.30) boils down to:

$$\text{Ent}_\mu(h) \geq \frac{\lambda}{2} W_2^2(h\,d\mu, \mu) = \frac{\lambda}{2} T_2(h\,d\mu, \mu).$$

Hence, when $\lambda > 0$ we obtain $\mu \in T_2(2/\lambda)$. Since $g = 1$ is not compactly supported, an approximation argument as in [13] is needed.

Proof of Theorem 2.24. The idea is to take $h = 1$ in (2.30). The right way to do is to choose a sequence of compactly supported h_n tending to 1, write the following argument with h_n and let n to infinity at the very end. Since this would just make the presentation more messy, we prefer to take $h = 1$ directly in (2.30):

$$\text{Ent}_\mu(g) \leq -\frac{\lambda}{2} \int d(x, \exp_x(\nabla\theta))^2 g\,d\mu - \int \langle \nabla\theta, \nabla g\rangle \, d\mu$$

$$\leq \int \left(-\frac{\lambda}{2}|\nabla\theta|^2 + |\nabla\theta|\frac{|\nabla g|}{g} \right) g\,d\mu.$$

If $\lambda > 0$ then for all $a, b \in \mathbb{R}$, it holds $-\frac{\lambda a^2}{2} + ab \leq \frac{b^2}{2\lambda}$, so we immediately deduce

$$\text{Ent}_\mu(g) \leq \frac{1}{2\lambda} \int \frac{|\nabla g|^2}{g^2} g\,d\mu,$$

that is Theorem 2.24(a). On the other hand, if $\lambda = -K \leq 0$, we have

$$\text{Ent}_\mu(g) \leq \frac{K}{2} \int d^2(x, T(x))g\,d\mu + \int |\nabla g| d(x, T(x))\,d\mu.$$

We upper-bound the last term in the previous expression by

$$\frac{1}{2\eta} \int \frac{|\nabla g|^2}{g}\,d\mu + \frac{\eta}{2} \int d^2(x, T(x))g\,d\mu,$$

where $\eta > 0$ is a small constant which will be chosen later. Therefore, we are left to control $\frac{K+\eta}{2} \int d^2(x, T(x))g\,d\mu$. Using that

$$d^2(x, T(x)) \leq (1+\eta)d^2(x, x_0) + \left(1 + \frac{1}{\eta}\right) d^2(x_0, T(x)),$$

we obtain

$$\frac{K+\eta}{2}\int d^2(x, T(x))g\,d\mu \leq \frac{K+\eta}{2}(1+\eta)\int d^2(x, x_0)g\,d\mu$$

$$+ \frac{K+\eta}{2}\left(1+\frac{1}{\eta}\right)\int d^2(x_0, T(x))g\,d\mu$$

$$= \frac{K+\eta}{2}(1+\eta)\int d^2(x, x_0)g\,d\mu$$

$$+ \frac{K+\eta}{2}\left(1+\frac{1}{\eta}\right)\int d^2(x_0, y)\,d\mu$$

$$\leq \frac{K+\eta}{2}(1+\eta)\int d^2(x, x_0)g\,d\mu + C(\eta),$$

where at the last step we used the integrability condition on μ. We now use the duality formula for the entropy (4.1) to deduce

$$(1-\eta)\int \frac{f}{1-\eta}g\,d\mu \leq (1-\eta)\left(\text{Ent}_\mu(g) + \log\left(\int e^{\frac{f}{1-\eta}}\,d\mu\right)\right).$$

Applying the above inequality with $f = \frac{K+\eta}{2}(1+\eta)d(x, x_0)^2$ and combining all together, we finally get

$$\text{Ent}_\mu(g) \leq (1-\eta)\text{Ent}_\mu(g) + (1-\eta)\log\left(\int e^{\frac{K+\eta}{2}\frac{1+\eta}{1-\eta}d(x,x_0)^2}\,d\mu\right)$$

$$+ C(\eta)\int \frac{|\nabla g|^2}{g}\,d\mu + C(\eta),$$

for all $g \geq 0$ such that $\int g\,d\mu = 1$. Hence, choosing η small enough so that $(K+\eta)\frac{1+\eta}{1-\eta} \leq K + \varepsilon$, we have

$$\text{Ent}_\mu(g) \leq C + D\int \frac{|\nabla g|^2}{g}\,d\mu,$$

for all $g \geq 0$ with $\int g\,d\mu = 1$. By homogeneity we get

$$\text{Ent}_\mu(g) \leq C\int g\,d\mu + D\int \frac{|\nabla g|^2}{g}\,d\mu,$$

for all $g \geq 0$. Using tightening techniques one can take $C = 0$ up to sufficiently increase D (see [4]), and this completes the proof of Theorem 2.24(b).   □

3 The Gaussian correlation conjecture

Is it true that, given two convex sets $A, B \subset \mathbb{R}^n$, both symmetric with respect to the origin, one has

$$\gamma_n(A \cap B) \geq \gamma_n(A)\gamma_n(B) ? \tag{3.1}$$

This is one of the most challenging questions about Gaussian measures. A positive answer would have many applications about suprema of Gaussian processes or small-ball probability estimates in Banach spaces. We start with recalling a functional version of the conjecture.

Definition 3.1. A non-negative function $\rho : \mathbb{R}^n \rightarrow [0, +\infty)$ is log-concave if, for all $\lambda \in (0, 1)$, for all $x, y \in \mathbb{R}^n$,

$$\rho(\lambda x + (1 - \lambda)y) \geq \rho(x)^\lambda \rho(y)^{1-\lambda}.$$

We remark that such a ρ has convex super-level sets: the sets $\{\rho \geq t\}$ are convex for all $t \geq 0$. Moreover any such ρ can be represented by a monotone limit of functions of the form $\sum_i \alpha_i \chi_{A_i}$, where $\alpha_i \geq 0$ and the sets A_i are convex. The conjecture, which can be written as

$$\int \chi_A \chi_B \, d\gamma_n \geq \int \chi_A \, d\gamma_n \int \chi_B \, d\gamma_n$$

for all A, B symmetric and convex, is therefore easily seen to be equivalent to

$$\int fg \, d\gamma_n \geq \int f \, d\gamma_n \int g \, d\gamma_n$$

for all f, g even and log-concave.

The correlation conjecture has been confirmed in a few particular cases. Inequality (3.1) is trivially true for $n = 1$, while for $n = 2$ it was proved by Pitt [39]. Several other cases have been treated, *e.g.* in the paper [41]. In the rest of this section we present some partial results in the direction of the conjecture, which have nice connexions with optimal transport.

3.1 The case when A is a strip

The following result holds:

Lemma 3.2 (Khatri [24], Sidak [42]). *The correlation conjecture is true if A is a symmetric strip.*

Before proving this result, let us recall the Brunn-Minkowski inequality and some useful consequences (see *e.g.* [38]): given two open sets $A, B \subset \mathbb{R}^n$ we have

$$|A + B| \geq \left(|A|^{1/n} + |B|^{1/n}\right)^n,$$

where $A + B$ denotes the Minkowski's sum of sets: $A + B := \{x + y : x \in A, y \in B\}$ and $|A| = \mathcal{L}^n(A)$. Combining this inequality with the arithmetic-geometric inequality, we get

$$|\lambda A + (1 - \lambda)B| \geq \left(\lambda|A|^{1/n} + (1 - \lambda)|B|^{1/n}\right)^n \geq |A|^\lambda |B|^{1-\lambda}$$

for all $\lambda \in [0, 1]$. On the other hand, one can prove that the the inequality

$$|\lambda A + (1 - \lambda)B| \geq |A|^\lambda |B|^{1-\lambda} \tag{3.2}$$

implies the Brunn-Minkowski Inequality. We remark that one main feature of the inequality (3.2) is to be dimension-independent. Its functional version is the Prékopa-Leindler inequality: let $f, g, h : \mathbb{R}^n \to [0, +\infty]$ be such that, for some $\lambda \in (0, 1)$,

$$h(\lambda x + (1 - \lambda)y) \geq f^\lambda(x)g^{1-\lambda}(y) \qquad \forall x, y \in \mathbb{R}^n.$$

Then

$$\int h \, dx \geq \left(\int f \, dx\right)^\lambda \left(\int g \, dx\right)^{1-\lambda}.$$

Observe that, by considering $h = \chi_{\lambda A + (1-\lambda)B}$, $f = \chi_A$ and $g = \chi_B$, we recover (3.2).

As another example, let ρ be a non-negative function and consider $h = \rho\chi_{\lambda A + (1-\lambda)B}$, $f = \rho\chi_A$ and $g = \rho\chi_B$. Then the hypotheses of the Prékopa-Leindler Inequality are satisfied if ρ is log-concave. Since the Gaussian density is log-concave it follows in particular that

$$\gamma_n(\lambda A + (1 - \lambda)B) \geq \gamma_n(A)^\lambda \gamma_n(B)^{1-\lambda} \qquad \forall \lambda \in [0, 1]. \tag{3.3}$$

We now prove the Khatri-Sidak Lemma. Without loss of generality we write points in $\mathbb{R}^n$ as (t, y) with $t \in \mathbb{R}$ and $y \in \mathbb{R}^{n-1}$, and we assume $A = [-a, a] \times \mathbb{R}^{n-1}$. We now recall a useful fact on sections of convex sets. If $K \subset \mathbb{R}^n$, we define for $t \in \mathbb{R}$, the section $K_t := \{y \in \mathbb{R}^{n-1} : (t, y) \in K\}$. If K is convex, then for all $s, t \in \mathbb{R}$ and $\lambda \in (0, 1)$ it holds

$$\lambda K_s + (1 - \lambda)K_t \subset K_{\lambda s + (1-\lambda)t}.$$

Combining this inclusion with Inequality (3.3) shows that

$$t \mapsto \gamma_{n-1}(K_t)$$

is log-concave. In particular, this function has convex upper-level sets. If K is convex and symmetric, this function is also even and its upper-level sets are symmetric intervals. We have

$$\gamma_n(A \cap B) = \int_{-a}^{a} \gamma_{n-1}(B_t) \, d\gamma_1(t)$$

$$= \int_0^1 \gamma_1\big(\{t \in [-a, a] : \gamma_{n-1}(B_t) > u\}\big) \, du.$$

But by the 1-dimensional version of the correlation conjecture we have

$$\gamma_1([-a, a] \cap [-b, b]) \geq \gamma_1([-a, a])\gamma_1([-b, b]) \qquad \forall b \geq 0,$$

and so

$$\gamma_n(A \cap B) \geq \gamma_1([-a, a]) \int_0^1 \gamma_1\big(\{t : \gamma_{n-1}(B_t) > u\}\big) \, du$$

$$= \gamma_1([-a, a])\gamma_n(B) = \gamma_n(A)\gamma_n(B),$$

which is the desired result. □

3.2   The case when A is an ellipsoid

Theorem 3.3 (Hargé [21]). *The conjecture is true if A is an ellipsoid.*

We will now show a proof of the above result given by Cordero-Erausquin [13] using optimal transport. The key ingredient is the following result:

Theorem 3.4 (Caffarelli [11]). *Let γ be a Gaussian probability measure on $\mathbb{R}^n$, and $\mu = \rho \, d\gamma$ be a probability measure with a log-concave density ρ. Then the optimal transport map T from γ to μ is 1-Lipschitz.*

Let us briefly sketch the idea of the proof and refer to [11] for a rigorous argument. We recall that the optimal transport map can be written as $\nabla\phi$, with ϕ convex. We start from the Monge-Ampère equation

$$\det(\nabla^2\phi(x)) = \frac{f(x)}{g(\nabla\phi(x))},$$

which, with the notation $\rho = e^{-V}$, can be equivalently written as

$$\log\big(\det(\nabla^2\phi(x))\big) = \log f(x) - \log g(\nabla\phi(x))$$

$$= -\frac{|x|^2}{2} + \frac{|\nabla\phi(x)|^2}{2} + V(\nabla\phi(x)), \tag{3.4}$$

with V convex. We now fix a point $x_0 \in \mathbb{R}^n$ where the maximum eigenvalue of $\nabla^2\phi(x)$ attains its maximum. Without loss of generality we can assume that the corresponding eigenvector is e_1, so that the maximum eigenvalue is given by $\partial_{11}\phi(x_0)$. Then we have

$$\nabla\partial_{11}\phi(x_0) = 0, \qquad \nabla^2\partial_{11}\phi(x_0) \leq 0.$$

Differentiating twice the equation at x_0 in the direction e_1 and exploiting the above equation together with the fact that $\nabla^2\phi(x_0)$ is nonnegative definite (and thus the same holds for $[\nabla^2\phi(x_0)]^{-1}$) one easily gets

$$\partial_{11}^2 \log\left(\det(\nabla^2\phi(x))\right)|_{x=x_0} = \operatorname{tr}\left([\nabla^2\phi(x_0)]^{-1} \cdot \nabla^2\partial_{11}^2\phi(x_0)\right)$$
$$- \left|[\nabla^2\phi(x_0)]^{-1} \cdot \nabla^2\partial_1\phi(x_0)\right|^2 \leq 0.$$

On the other hand, by differentiating the right hand side in (3.4) one gets

$$\partial_{11}^2 \log\left(\det(\nabla^2\phi(x))\right)|_{x=x_0} = -1 + |\nabla\partial_1\phi|^2(x_0)$$
$$+ D^2V(\nabla\phi(x_0))\left[\nabla\partial_1\phi(x_0), \nabla\partial_1\phi(x_0)\right]$$
$$\geq -1 + |\nabla\partial_1\phi|^2(x_0).$$

Comparing the two expressions we get $\partial_{11}^2\phi(x_0) \leq 1$. Hence the eigenvalues of $\nabla^2\phi$ are everywhere bounded from above by 1, which proves that $T = \nabla\phi$ is 1-Lipschitz.   □

We now prove Theorem 3.3. Up to a linear change of variables (which transforms the standard Gaussian measure γ_n into a different Gaussian measure γ) we can assume that $A = B^n$ is the Euclidean unit ball, and we want to show that

$$\gamma(B^n \cap K) \geq \gamma(B^n)\gamma(K)$$

for any symmetric convex set K, or equivalently

$$\mu(B^n) \geq \gamma(B^n),$$

where $\mu = \frac{\chi_K}{\gamma(K)}\gamma$. Since $\frac{\chi_K}{\gamma(K)}$ is log-concave, we can apply Theorem 3.4 to obtain that the optimal map T from γ to μ is a contraction. We now observe that, by symmetry, also the map $x \mapsto -T(-x)$ is an optimal map, so by uniqueness $T(x) = -T(-x)$. In particular $T(0) = 0$. This implies $T(B^n) \subset B^n$, and so

$$\mu(B^n) = \gamma(T^{-1}(B^n)) \geq \gamma(B^n),$$

as wanted.   □

A slab being a degenerate ellipsoid, the Khatri-Sidak lemma is a limit case of Hargé's result on ellipsoids. The proof we gave for the Khatri-Sidak lemma could indeed be interpreted as a transportation proof based on the Knothe map [25], which can be viewed as a limit of optimal transports for degenerating Euclidean distances.

3.3 Correlation for a convex and a log-concave function

Let f be a even log-concave function, and g a even convex function. Then $e^{-\varepsilon g}$ is even and log-concave for any $\varepsilon > 0$. Assuming the Gaussian Correlation Conjecture to be true, we would get

$$\int f e^{-\varepsilon g} \, d\gamma_n \geq \int f \, d\gamma_n \int e^{-\varepsilon g} \, d\gamma_n,$$

which implies, by a first order Taylor expansion at $\varepsilon = 0$,

$$\int f g \, d\gamma_n \leq \int f \, d\gamma_n \int g \, d\gamma_n.$$

It turns out that the above inequality is indeed true, as proved by Hargé [22]. The aim of this subsection is to present his argument. We have to show that

$$\int g \frac{f}{\int f \, d\gamma_n} \, d\gamma_n \leq \int g \, d\gamma_n,$$

with f even and log-concave, g even and convex and, with no loss of generality, we can assume $\int f \, d\gamma_n = 1$. Let T be the optimal transport map from γ_n to $\frac{f}{\int f \, d\gamma_n} \gamma_n$. We recall that $T = \nabla \varphi$ with φ convex and, by Theorem 3.4, Hess $\varphi \leq I_n$. Let us define $\psi(x) := |x|^2/2 - \varphi(x)$. Then

$$T(x) = x - \nabla \psi(x), \qquad 0 \leq \text{Hess } \psi \leq I_n.$$

We therefore have

$$\int g f \, d\gamma_n = \int g \circ T \, d\gamma_n = \int g(x - \nabla \psi(x)) \, d\gamma_n(x),$$

and we need to prove that

$$\int g(x - \nabla \psi(x)) \, d\gamma_n(x) \leq \int g \, d\gamma_n. \qquad (3.5)$$

Recall that the Ornstein-Uhlenbeck semigroup

$$P_t = e^{tL}, \qquad L = \Delta - x \cdot \nabla,$$

is explicitly given, for good enough functions $f : \mathbb{R}^n \to \mathbb{R}$, by

$$P_t f(x) := \int f(e^{-t} x + \sqrt{1 - e^{-2t}} \, y) \, d\gamma_n(y).$$

It satisfies $P_0 f = f$, $\lim_\infty P_t f = \int f \, d\gamma_n$ and the integration by parts formula

$$\int f L g \, d\gamma_n = - \int \nabla f \cdot \nabla g \, d\gamma_n. \tag{3.6}$$

It is convenient to extend the action of the semigroup to vector-valued and matrix-valued functions (by action of P_t on coefficients). Set

$$\theta(t) = \int g(x - P_t \nabla \psi(x)) \, d\gamma_n(x).$$

Then

$$\theta(0) = \int g(x - \nabla \psi(x)) \, d\gamma_n(x),$$

while

$$\lim_{t \to +\infty} \theta(t) = \int g\left(x - \int \nabla \psi \, d\gamma_n\right) d\gamma_n(x) = \int g \, d\gamma_n,$$

where we used that

$$\int \nabla \psi \, d\gamma_n = \int (\nabla \psi(x) - x) \, d\gamma_n(x) = - \int T(x) \, d\gamma_n(x)$$
$$= - \int y f(y) \, d\gamma_n(y) = 0,$$

as f is even. Hence to prove (3.5) it suffices to show that θ is non-decreasing. Let us compute its derivative: using that $\partial_t P_t f = L P_t f$ and a tedious integration by parts in the form of (3.6) with $V(x) = |x|^2/2$, we get

$$\theta'(t) = - \int \nabla g(x - P_t \nabla \psi(x)) \cdot L(P_t \nabla \psi) \, d\gamma_n = \int \text{tr}(H(I - M^t)M) \, d\gamma_n(x)$$

where $H(x) = \text{Hess } g(x - P_t \nabla \psi(x))$ and $M(x) = \nabla(P_t \nabla \psi)(x)$. Since $\nabla P_t f = e^{-t} P_t \nabla f$ for any function f, we get

$$M = M^t = e^{-t} P_t \text{ Hess } \psi,$$

which is definite positive with eigenvalues bounded by 1. Thus all matrices $H, M, I - M$ are non-negative definite (recall that g is convex), which implies

$$\text{tr}\left(H(I - M^t)M\right) = \text{tr}\left(\sqrt{M(I - M)} H \sqrt{M(I - M)}\right) \geq 0$$

thanks to the identity $\text{tr}(AB) = \text{tr}(BA)$ with $A = H\sqrt{(I - M^t)M}$ and $B = \sqrt{(I - M^t)M}$. Hence $\theta'(t) \geq 0$ for all $t \geq 0$, and the proof is complete. □

4 Appendix: basic results on the Relative Entropy functional

A basic duality formula for the Entropy functional is:

$$\text{Ent}_\mu(f) = \sup\left\{\int fg\,d\mu : g \in C_b, \int e^g\,d\mu \leq 1\right\}. \qquad (4.1)$$

For the Relative Entropy the inequality reads as follows:

$$\text{Ent}(\nu|\mu) = \sup\left\{\int g\,d\nu : g \in C_b, \int e^g\,d\mu \leq 1\right\}. \qquad (4.2)$$

The proof can be found *e.g.* in [2]: it uses the elementary inequality $uv \leq u\ln u - u + e^v$, for $u \geq 0$ and $v \in \mathbb{R}$. Assuming with no loss of generality (because both sides are positively 1-homogeneous) $\int f\,d\mu = 1$ we can use it with $u = f(x)$ and $v = g(x)$ to get

$$\int f\ln f\,d\mu \geq \int fg\,d\mu + 1 - \int e^g\,d\mu,$$

and this proves the inequality $\geq$. Simple consequences of the duality formula are:

– the joint lower semicontinuity of relative entropy, namely

$$\text{Ent}(\nu|\mu) \leq \liminf_{n\to\infty}\text{Ent}(\nu_n|\mu_n)$$

whenever $\mu_n \to \mu$ and $\nu_n \to \nu$ in the duality with C_b;
– the marginal monotonicity, namely

$$\text{Ent}(p_\sharp\nu|p_\sharp\mu) \leq \text{Ent}(\nu|\mu)$$

whenever $p : X \to Z$ is a Borel map. We actually need a more precise version of the marginal monotonicity. We state and prove it in a special case, when X is a product space and p is the projection on the first factor.

Lemma 4.1 (Entropy additivity). *Let $\theta, \lambda \in \mathscr{P}(X \times Y)$, let $\sigma, \mu \in \mathscr{P}(X)$ be their respective marginals and set $d\theta(x, y) = d\theta_x(y)d\sigma(x)$, $d\lambda(x, y) = d\lambda_x(y)d\mu(x)$. Then*

$$\text{Ent}(\theta|\lambda) = \text{Ent}(\sigma|\mu) + \int_X \text{Ent}(\theta_x|\lambda_x)\,d\sigma^a(x) \qquad (4.3)$$

where $\sigma = \sigma^a + \sigma^s$ is the Radon-Nikodym decomposition of σ in absolutely continuous and singular part with respect to μ.

A technical point: the conditional measure λ_x is well defined only up to μ-negligible sets. It is for this reason that only σ^a appears in (4.3), to make the equality always meaningful. But if $\sigma^s \neq 0$ then both $\text{Ent}(\sigma | \mu)$ and $\text{Ent}(\theta | \lambda)$ are equal to $+\infty$. So, in the sequel we can always assume $\sigma \ll \mu$.

In order to prove the lemma, we consider first the case when θ is not absolutely continuous with respect to λ, so that there exists a Borel set A such that $\lambda(A) = 0$ and $\theta(A) > 0$. In this case the left hand side equals $+\infty$. If either σ is not absolutely continuous with respect to μ or θ_x is not absolutely continuous with respect to λ_x for σ^a-a.e. x we are done (both sides are $+\infty$), otherwise setting $A_x = \{y : (x, y) \in A\}$ from $\int \lambda_x(A_x) \, d\mu = \lambda(A) = 0$ we deduce that $\lambda_x(A_x) = 0$ μ-a.e. But then, since $\int \theta_x(A_x) \, d\sigma = \theta(A) > 0$, it must be $\theta_x(A_x) > 0$ in a set of strictly positive σ-measure. This set must also have strictly positive μ-measure, contradicting the fact that $\theta_x \ll \lambda_x$ for μ-a.e. x.

So, in the sequel we can assume $\theta = f\lambda$ and we can define

$$Z(x) := \int f(x, y) \, d\lambda_x(y), \tag{4.4}$$

so that $\sigma = Z\mu$ and

$$\text{Ent}(\sigma | \mu) = \int Z \ln Z \, d\mu(x). \tag{4.5}$$

Writing

$$d\theta(x, y) = df(x, \cdot)\lambda_x(y)d\mu(x) = \frac{1}{Z(x)} f(x, \cdot)d\lambda_x(y)d\sigma(x)$$

we see that $\theta_x = Z^{-1}(x)f(x, \cdot)\lambda_x$ for σ-a.e. x and therefore for μ-a.e. x with $Z(x) \neq 0$. So we can write

$$\int_X \text{Ent}(\theta_x | \lambda_x) \, d\sigma(x)$$

$$= \int_{\{Z>0\}} \int \frac{f(x, y)}{Z(x)} \ln \frac{f(x, y)}{Z(x)} \, d\lambda_x(y)Z(x)d\mu(x) \tag{4.6}$$

$$= \int f \ln f \, d\lambda - \int Z \ln Z d\mu$$

that, combined with (4.5), gives (4.3).

When $\lambda = \mu \times \nu$ we can also read (4.6), taking the equality $\lambda_x = \nu$ and (4.4) into account, in the simpler and expressive form

$$\text{Ent}_{\mu \times \nu}(f) = \text{Ent}_\mu \left(\int_Y f(\cdot, y) \, d\nu(y) \right) + \int_X \text{Ent}_\nu(f(x, \cdot)) \, d\mu(x). \tag{4.7}$$

References

[1] L. AMBROSIO, N. GIGLI and G. SAVARÉ, "Gradient flows in metric spaces and in the space of probability measures" Second edition, lectures in Mathematics ETH Zürich, Birkhäuser Verlag, Basel, 2008.

[2] C. ANÉ, S. BLACHÈRE, D. CHAFAI, P. FOUGÈRES, I. GENTIL, F. MALRIEU, C. ROBERTO and G. SCHEFFER, *Sur les inégalités de Sobolev logarithmiques*, In: "Panoramas et Synthèses [Panoramas and Syntheses]" Vol. 10, Société Mathématique de France, Paris, 2000.

[3] D. BAKRY and M. EMERY, *Diffusions hypercontractives*, In: "Séminaire de Probabilités XIX", n. 1123 in Lect. Notes in Math., Springer, 1985, 179–206.

[4] F. BARTHE and A. KOLESNIKOV, *Mass transport and variants of the logarithmic Sobolev inequality*, J. Geom. Anal. **18** (2008), 921–979.

[5] F. BARTHE and C. ROBERTO, *Sobolev inequalities for probability measures on the real line*, Studia Math. **159** (2003), 481–497.

[6] S. G. BOBKOV, I. GENTIL and M. LEDOUX, *Hypercontractivity of Hamilton-Jacobi equations*, J. Math. Pures Appl. **80** (2001), 669–696.

[7] S. G. BOBKOV and F. GÖTZE, *Exponential integrability and transportation cost related to logarithmic Sobolev inequalities*, J. Funct. Anal. **163** (1999), 1–28.

[8] V. BOGACHEV, "Gaussian measures" (English summary) Mathematical Surveys and Monographs, 62, American Mathematical Society, Providence, RI, 1998.

[9] C. BORELL, *The Brunn-Minkowski inequality in Gauss space*, Invent. Math. **30** (1975), 207–216.

[10] Y. BRENIER, *Polar factorization and monotone rearrangement of vector-valued functions*, Comm. Pure Appl. Math **44** (1991), 375–417.

[11] L. CAFFARELLI, *Monotonicity properties of optimal transportation and the FKG and related inequalities*, Comm. Math. Phys. **214** (2000), 547–563.

[12] P. CATTIAUX and A. GUILLIN, *On quadratic transportation cost inequalities*, J. Math. Pures Appl. (9) **86** (2006), 341–361.

[13] D. CORDERO-ERAUSQUIN, *Some applications of mass transport to Gaussian-type inequalities*, Arch. Rational Mech. Anal. **161** (2000), 257–269.

[14] D. CORDERO-ERAUSQUIN, R. J. MCCANN and M. SCHMUCK-ENSCHL GER, *Prékopa-Leindler type inequalities on Riemannian manifolds, Jacobi fields, and optimal transport*, Ann. Fac. Sci. Toulouse Math. (6) **15** (2006), 613–635.

[15] J. D. DEUSCHEL and D. W. STROOCK, "Large deviations", of Pure and Applied Mathematics, Vol. 137, Academic Press Inc., 1989.

[16] A. DEMBO and O. ZEITOUNI, "Large deviation techniques and applications, second edition", Applications of Mathematics 38, Springer Verlag, 1998.

[17] L. C. EVANS, *Partial Differential Equations*, Graduate Studies in Mathematics American Mathematical Society **19** (1998).

[18] N. GOZLAN, *Characterization of Talagrand's like transportation-cost inequalities on the real line*, J. Funct. Anal. **250** (2007), 400–425.

[19] N. GOZLAN, *Characterization of dimension free concentration in terms of transportation inequalities*, preprint, arXiv:0804.3089.

[20] N. GOZLAN, *Poincaré inequalities and dimension free concentration of measure*, to appear in Annales Inst. H. Poincaré, Probab. Statist., to appear.

[21] G. HARGÉ, *A particular case of correlation inequality for the Gaussian measure*, Ann. Probab. **27** (1999), 1939–1951.

[22] G. HARGÉ, *A convex/log-concave correlation inequality for Gaussian measure and an application to abstract Wiener spaces*, Probab. Theory Relat. Fields **130** (2004), 415–440.

[23] W. B. JOHNSON and J. LINDENSTRAUSS, *Extension of Lipschitz mappings into a Hilbert space*, Contemp. Math. **26** (1984), 189–206.

[24] C. G. KHATRI, *On certain inequalities for normal distributions and their applications to simultaneous confidence bounds*, Ann. Math. Statist. **38** (1967), 1853–1867.

[25] H. KNOTHE, *Contributions to the theory of convex bodies*, Michigan Math. J. **4** (1957), 39–52.

[26] M. LEDOUX, *Concentration of measure and logarithmic Sobolev inequalities*, In: "Séminaire de Probabilités, XXXIII" in Lecture Notes in Math., n. 1709, Springer, Berlin, 1999, 120–216.

[27] M. LEDOUX, "The concentration of measure phenomenon", of Mathematical Surveys and Monographs, American Mathematical Society, Vol. 89, Providence, RI, 2001.

[28] M. LEDOUX, *Isoperimetry and Gaussian analysis*, "Lectures on probability theory and statistics" (Saint-Flour, 1994), Lecture Notes in Math., n. 1648, Springer, 1996, 165–294.

[29] R. J. MCCANN, *A convexity principle for interacting gases*, Adv. Math. **128** (1997), 153–179.

[30] R. J. MCCANN, *Existence and uniqueness of monotone measure-preserving maps*, Duke Math. J. **80** (1995), 309–323.

[31] R. J. MCCANN, *Polar factorization of maps on Riemannian manifolds*, Geom. Funct. Anal. **11** (2001), 589–608.

[32] K. MARTON, *Bounding d̄-distance by informational divergence*, Ann. Probab. **24** (1996), 857–866.

[33] J. MATOUŠEK, "Lectures on discrete geometry", Graduate Texts in Mathematics, n. 212, Springer-Verlag, New York, 2002.

[34] V. MILMAN and G. SCHECHTMAN, *Asymptotic Theory of Finite Dimensional Normed Spaces*, In: "Lecture Notes" in Math. n. 1200, Springer Verlag, 1986.

[35] L. MICLO and C. ROBERTO, "In égalités de Hardy et de Sobolev logarithmique", Thèse de Doctorat de C. Roberto, chapter 3, Université Paul Sabatier, 2001.

[36] B. MUCKENHOUPT, *Hardy's inequality with weights* Studia Math. **44** (1972), 31–38.

[37] F. OTTO and C. VILLANI *Generalization of an inequality by Talagrand and links with the logarithmic Sobolev inequality*, J. Funct. Anal. **173** (2000), 361–400.

[38] G. PISIER, "The volume of convex bodies and Banach space geometry" Cambridge Tracts in Mathematics, n. 94, Cambridge University Press, 1989.

[39] L. PITT, *A gaussian correlation inequality for symmetric convex sets*, Ann. Probab. **5** (1997), 470–474.

[40] S. T. RACHEV and L. RÜSCHENDORF, "Mass Transportation Problems", Probab. Appl. Springer-Verlag, 1998.

[41] G. SCHECHTMAN, TH. SCHLUMPRECHT and J. ZINN, *On the Gaussian measure of the intersection*, Ann. Probab. **26** (1998), 346–357.

[42] Z. SIDAK, *Rectangular confidence regions for the means of multivariate normal distributions*, J. Amer. Stat. Assoc. **62** (1967), 626–633.

[43] V. N. SUDAKOV and B. S. TSIREL'SON, *Extremal propreties of half-spaces for spherically invariant measures*, J. Soviet Math. **9** (1978), 9–18. Translated from Zap. Nauchn. Sem. Leningrad. Otdel. Math. Inst. Steklova. **41** (1974), 14–24.

[44] M. TALAGRAND, *Transportation cost for Gaussian and other product measures*, Geom. Funct. Anal. **6** (1996), 587–600.

[45] G. TALENTI, *Osservazioni sopra una classe di disuguaglianze*, Rend. Sem. Mat. Fis. Milano **39** (1969), 171–185.

[46] G. TOMASELLI, *A class of inequalities*, Bull. Un. Mat. Ital. **21** (1969), 622–631.

[47] C. VILLANI, "Optimal transport, old and new", Grundelehren der mathematischen Wissenschaften n. 338, Springer, 2008.

[48] C. VILLANI, *Topics in optimal transportation*, of "Graduate Studies in Mathematics", American Mathematical Society, Vol. 58, Providence, RI, 2003.

[49] F.-Y. WANG, *Logarithmic Sobolev inequalities on noncompact Riemannian manifolds*, Proba. Theory Relat. Fields **109** (1997), 417–424.

A weak KAM theorem;
from finite to infinite dimension

Wilfrid Gangbo and Adrian Tudorascu

Abstract. These notes contain a series of lectures given by the first author in the 2008 GNAMPA–INDAM School in Pisa. It is based on recent results by both authors who initiated a study of an infinite dimensional weak KAM theory. While some of the results presented here have already appeared in their joint work [16], the core of this manuscript, section 3, has never been submitted for publication anywhere. The current manuscript should be regarded as a companion to [16].

1 Introduction

To understand the subtlety of the so-called cell problem appearing in the KAM and the Weak KAM theories, let us start with a very simple Hamiltonian. Consider the one-dimensional Hamiltonian $h(x, p) = |p|^2 - \sin^2(\pi x)$ for $x, p \in \mathbb{R}$. As done in the *KAM theory* let us proceed with the exercise of searching for real numbers λ and functions $s \in C^1(\mathbb{R})$ such that s' is $\mathbb{Z}$-periodic and satisfies the equation $h(x, s'(x)) = \lambda$. Since every function $s \in C^1(\mathbb{R})$ such that s' is $\mathbb{Z}$-periodic can be decomposed as the sum of a $\mathbb{Z}$-periodic function $u \in C^1(\mathbb{R})$ and a linear function $x \to cx$, the problem at hand is to find two reals numbers λ, c and a $\mathbb{Z}$-periodic function $u \in C^1(\mathbb{R})$ such that

$$h(x, c + u'(x)) = \lambda \tag{1.1}$$

has a solution. For such a simple Hamiltonian, at a first glance, (1.1) might look like a simple problem. It is wrong to think that existence of solutions depends mainly on whether or not λ is in the range of h. If $c = 0$

The first author gratefully acknowledges the support provided by NSF grants DMS-02-00267, DMS-03-54729 and DMS-06-00791.

The second author gratefully acknowledges the support provided by the School of Mathematics at Georgia Tech.

or $|c| = 2/\pi$ one readily checks that unless $\lambda = 0$ there is no $\mathbb{Z}$-periodic function $u \in C^1(\mathbb{R})$ satisfying (1.1). If $0 < |c| < 2/\pi$ then (1.1) has no $\mathbb{Z}$–periodic solution $u \in C^1(\mathbb{R})$. In the *Weak KAM theory* one seeks for real numbers λ and $\mathbb{Z}$-periodic Lipschitz functions $u \in C(\mathbb{R})$ viscosity solutions of (1.1). For $c \in \mathbb{R}$ prescribed it is well-known that (1.1) admits a periodic viscosity solution for exactly one value of λ denoted $\bar{h}(c)$. This is one way of defining the effective Hamiltonian $\bar{h}$ which can be explicitly computed as a function of c [18]. The so-called cell problem (1.1) plays a central role in understanding the dynamics of the Hamiltonian flow of h. Especially as we are dealing with a Hamiltonian defined on the cotangent bundle to a one-dimensional manifold, (1.1) can be used to identify periodic solutions of the flow associated h (*cf.* [22]).

The purpose of [16] has been to extend techniques of the Weak KAM theory – proven to be a powerful tool for understanding finite dimensional Hamiltonian systems – to some classes of PDEs viewed as infinite dimensional Hamiltonian systems. In these notes, we show that the infinite dimensional analogue of (1.1) can be obtained by studying finite dimensional Hamiltonian systems. In order to explain the link between the existing (finite dimensional) Weak KAM theory and our approach let us fix a $\mathbb{Z}$-periodic function $W \in C^2(\mathbb{R})$. For each positive integer n consider the evolutive system

$$
\begin{cases}
\dot{x}_i = np_i \\
\dot{p}_i = -\dfrac{1}{n^2} \sum_{j=1}^{n} W'(x_i - x_j) \\
x_i(0) = \bar{x}_i, \quad p_i(0) = \bar{p}_i.
\end{cases}
\tag{1.2}
$$

Here, $\bar{x}, \bar{p} \in \mathbb{R}^n$ are prescribed. Clearly, (1.2) is a Hamiltonian system where the Hamiltonian and the Lagrangian are defined on $\mathbb{R}^n \times \mathbb{R}^n$ by

$$
h(x, p) = \frac{n}{2}|p|^2 + \frac{1}{2n^2} \sum_{i,j=1}^{n} W(x_i - x_j),
$$

$$
l(x, v) = \frac{1}{2n}|v|^2 + \frac{1}{2n^2} \sum_{i,j=1}^{n} W(x_i - x_j).
\tag{1.3}
$$

Let $(\bar{x}, \bar{p})$ be the initial conditions in (1.2). The Hamiltonian flow is ϕ defined by $\phi_t(\bar{x}, \bar{p}) = (x(t), p(t))$. Let $\mathcal{G}_n$ be the group of permutations of n letters. For $\tau \in \mathcal{G}_n$ and $k \in \mathbb{Z}^n$ we define

$$
P_\tau(x, p) = (x^\tau, p^\tau), \quad T_k(x, p) = (x + k, p)
$$

where $x, p \in \mathbb{R}^n$ and $x^\tau := (x_{\tau(1)}, \cdots, x_{\tau(n)})$. Since $h \circ P_\tau = h = h \circ T_k$ we have

$$\phi_t \circ P_\tau = \phi_t = \phi_t \circ T_k. \tag{1.4}$$

Thus, h can be viewed as a function defined on $(\mathcal{M}_n/\mathcal{G}_n) \times \mathbb{R}^n$ and ϕ can be viewed as a flow on the cotangent bundle of the n-symmetric product of the circle $\mathcal{M}_n/\mathcal{G}_n$, known to be a manifold [23]. Here, $\mathcal{M}_n$ is the n-dimensional flat torus and we shall identify the cotangent bundle of $\mathcal{M}_n/\mathcal{G}_n$ with $(\mathcal{M}_n/\mathcal{G}_n) \times \mathbb{R}^n$.

If a probability measure μ on the cotangent bundle $T^*\mathcal{M}_n$ is invariant under the flow ϕ, then

$$\int_{T^*\mathcal{M}_n} \langle \nabla u(x), \nabla_p h(x, p)\rangle \mathrm{d}\mu(x, p)$$
$$= \frac{\mathrm{d}}{\mathrm{d}t} \int_{T^*M} f(\phi_t(x, p))\mathrm{d}\mu(x, p)\Big|_{t=0} = 0, \tag{1.5}$$

where $f(x, p) = u(x)$ and u is continuous and $\mathbb{Z}^n$-periodic. We write $u \in C^1(\mathcal{M}_n)$. Let $F(x, p) = (x, \nabla_p h(x, p))$ be the Legendre map associated to h. The push-forward of μ by F is the measure $\nu := F_\# \mu$ defined on the tangent bundle $T\mathcal{M}_n$ by

$$\nu(B) = \mu\big(F^{-1}(B)\big)$$

for all Borel sets $B \subset T\mathcal{M}_n$. Note that (1.5) reads off

$$\int_{T\mathcal{M}_n} \langle \nabla u(x), v\rangle \mathrm{d}\nu(x, v) = 0. \tag{1.6}$$

One says that ν is weakly invariant under the flow ϕ, although the definition in (1.6) does not involve h or ϕ. Suppose $S \in C^2(\mathbb{R}^n)$ and ∇S is $\mathbb{Z}^n$-periodic. In other words, we are selecting a closed one-form $(x, \xi) \in \mathcal{M}_n \times \mathbb{R}^n \to \Lambda_x(\xi)$ of class C^1 on $\mathcal{M}_n$. One readily checks existence of a $\vec{c} \in \mathbb{R}^n$ (characterizing the cohomology class of Λ) such that

$$\int_{\mathcal{M}_n \times \mathbb{R}^n} \langle \nabla S(x), v\rangle \mathrm{d}\nu(x, v) = \langle \vec{c}, R(\nu)\rangle.$$

Here,

$$R(\nu) := \int_{T\mathcal{M}_n} v \mathrm{d}\nu(x, v)$$

and is referred to as the rotation number of ν.

In the Weak KAM theory one seeks for Borel measures on the tangent bundle $T\mathcal{M}_n$ that minimize the action

$$\nu \to \mathcal{A}_n(\nu) := \int_{T\mathcal{M}_n} l d\nu.$$

The minimization is performed over the set of weakly invariant measures ν of prescribed rotation vector $\vec{r} \in \mathbb{R}^n$. For a class of Lagrangians including those appearing in (1.3) such minimal measures are known to exist and are supported by the subdifferentials of functions $x \to \langle \vec{c}, x \rangle + u(x)$. Here $u \in C(\mathcal{M}_n)$ is a viscosity solutions of the cell problem

$$h(x, \vec{c} + \nabla u) = \bar{h}(\vec{c}) \tag{1.7}$$

and $\vec{c} \in \mathbb{R}^n$ is related to $\vec{r}$. We have denoted by $\bar{h}$ the effective Hamiltonian of h, defined by the fact that $\bar{h}(\vec{c})$ is the unique real number λ such that $h(x, \vec{c} + \nabla u) = \lambda$ admits a viscosity solution $u \in C(\mathcal{M}_n)$.

In these notes, not only are we interested in measures ν that minimize $\mathcal{A}_n$ over the set of weakly invariant measures of precribed rotation number $\vec{r}$, but we also require these measures to be invariant under the action of the group $\mathcal{G}_n : P_{\tau\#}\nu = \nu$ for all $\tau \in \mathcal{G}_n$. The latter condition yields that $\vec{r}$ must be parallel to $(1, \cdots, 1) \in \mathbb{R}^n$. It becomes natural to impose in (1.7) that $\vec{c}$ must be parallel to $(1, \cdots, 1) \in \mathbb{R}^n$. As a matter of fact, only for these special $\vec{c}$, were we able to show that as we let n tend to infinity, the finite dimensional solutions of (1.7) converge to their infinite dimensional analogue.

A formal explanation for restricting ourselves to $\vec{c}$ which are parallel to $(1, \cdots, 1) \in \mathbb{R}^n$ is based on the link between (1.2) and the Vlasov systems. The starting point is to view $T^*\mathcal{M}_n$ as a subset of $\mathcal{P}_2(\mathbb{R}^2)$, the set of Borel probability measures on $\mathbb{R}^2$ of bounded second moments. The embedding is given by $(\bar{x}, \bar{p}) \to 1/n \sum_{i=1}^n \delta_{(\bar{x}_i, \bar{p}_i)}$. Hence, to the path $t \to (x(t), p(t)) \in T^*\mathcal{M}_n$ satisfying (1.2) we associate the path $t \to f_t \in \mathcal{P}_2(\mathbb{R}^2)$ defined for each $t \geq 0$ by

$$f_t = \frac{1}{n} \sum_{i=1}^n \delta_{(x_i(t), \dot{x}_i(t))}.$$

Let ϱ_t be the first marginal of f_t: $\varrho_t = \frac{1}{n} \sum_{i=1}^n \delta_{x_i(t)}$ and set $P_t = \varrho_t * W$. The system of equations (1.2) translates into the so-called Vlasov system

$$\begin{cases} \partial_t f_t + v \partial_x f_t = \partial_x P_t \partial_v f_t \\ P_t(x) = \int_{\mathbb{R}} W(x - \bar{x}) d\rho_t(x) \\ f_0 = \bar{f} := \frac{1}{n} \sum_{i=1}^n \delta_{(\bar{x}_i, n\bar{p}_i)}. \end{cases} \tag{1.8}$$

The first equation in (1.8) must be understood in the sense of distributions. While (1.8) is a richer system than (1.2) in the sense that it encompasses the case $n = \infty$, both systems coincide when $n < \infty$. In the latter case both systems represent the evolution of n undistinguishable particles of same mass. The fact that the particles are undistinguishable explains why the rotation vectors of interest, from the point of view of the Vlasov systems, must be vectors whose components are equal.

Gangbo [27] has noticed that (1.8) can be regarded as an infinite-dimensional Hamiltonian ODE on the space of Borel probability measures on $\mathbb{R}^2$ with finite second-order moments (cf. also [1] and [14]). Indeed, if

$$\mathcal{H}(f) := \iint_{\mathbb{R}^2} \left[\frac{v^2}{2} + \frac{1}{2} \iint_{\mathbb{R}^2} W(x - y) df(y, w) \right] df(x, v),$$

then one may regard (1.8) as

$$\partial_t f + \mathrm{div}\left[J \nabla_w \mathcal{H}(f) f \right] = 0,$$

where J is the clockwise rotation matrix of angle $\pi/2$, and ∇_w is the Wasserstein gradient [1]. In the current manuscript we look for some special solutions, which allow for a connection with a more conventional way of regarding (1.8) as Hamiltonian. Assume the initial data is in the set of probabilities on $\mathbb{R}^2$ such that $f_0 = (M_0, N_0)_\# \nu_0$, where ν_0 is the Lebesgue measure on $(0, 1)$ and $M_0, N_0 \in L^2(0, 1)$. This means

$$\int_{\mathbb{R}^2} \varphi(x, v) df_0(x, v) = \int_0^1 \varphi(M_0(y), N_0(y)) dy \text{ for all } \varphi \in C_c(\mathbb{R}^2).$$

Let us introduce the initial value problem

$$\ddot{\sigma}_t z = -\int_0^1 W'(\sigma_t z - \sigma_t w) dw, \qquad \sigma_0 = M, \quad \dot{\sigma}_0 = N. \qquad (1.9)$$

This is an evolutive system on the infinite dimensional manifold $L^2(0, 1)$, which is a separable Hilbert space. We denote its inner product by $\langle \cdot, \cdot \rangle$ and its norm by $\| \cdot \|$. The space $L^2(0, 1)$ has a natural differential structure and at each $M \in L^2(0, 1)$ the tangent space at M is $T_M L^2(0, 1) = L^2(0, 1)$. Hence, the tangent bundle is $T L^2(0, 1) := L^2(0, 1) \times L^2(0, 1)$ which we identify with the cotangent bundle.

Let $L^2_{\mathbb{Z}}(0, 1)$ be the set of $M \in L^2(0, 1)$ whose ranges are contained in $\mathbb{Z}$. We define the $L^2_{\mathbb{Z}}(0, 1)$-torus by

$$\mathbb{T} := L^2(0, 1)/L^2_{\mathbb{Z}}(0, 1). \qquad (1.10)$$

We say that $\mathcal{W} : L^2(0, 1) \to \mathbb{R}$ is $L_{\mathbb{Z}}^2(0, 1)$-periodic if $\mathcal{W}(M + Z) = \mathcal{W}(M)$ for all $M \in L^2(0, 1)$ and all $Z \in L_{\mathbb{Z}}^2(I)$. We view $\mathcal{W}$ as a function defined on the $\mathbb{T}$. If, in addition, $\mathcal{W}$ is continuous, we write $\mathcal{W} \in C(\mathbb{T})$.

Suppose Λ is a $L_{\mathbb{Z}}^2(0, 1)$-periodic, differentiable, closed one-form on $L^2(0, 1)$ in the sense of [16, Section 5]. Suppose that $M \to \Lambda_M(M)$ is Lipschitz and rearrangement invariant and $L_{\mathbb{Z}}^2(0, 1)$-periodic. Suppose the second moment of γ, a Borel probability on $TL^2(0, 1)$, is finite:

$$\int_{TL^2(0,1)} l_2(N)d\gamma(M, N) < \infty, \qquad l_2(N) := \|N\|_{L^2(0,1)}^2.$$

By a *rearrangement invariant* map U defined on $L^2(0, 1)$ we understand a map satisfying $U(M) = U(N)$ for all $M, N \in L^2(0, 1)$ such that $M_\# v_0 = N_\# v_0$. Then there exists a real number c and a Lipschitz function $U \in C^1(\mathbb{T})$ such that

$$\Lambda_M(N) = c + dU_M(N)$$

for $M, N \in L^2(0, 1)$. If γ is a Borel measure on $TL^2(0, 1)$ invariant under the flow Ψ in the sense that $\Psi_{t\#}\gamma = \gamma$ for all $t > 0$ we use arguments similar to those appearing in (1.5) to obtain that

$$\int_{TL^2(0,1)} \Lambda_M(N)d\gamma(M, N) = R(\gamma)\, c,$$

where

$$R(\gamma) := \int_{TL^2(0,1)} l(N)d\gamma(M, N) \qquad l(N) := \int_0^1 N dv_0.$$

We refer to $R(\gamma)$ as the *rotation number* of γ.

If $W \in C^{1,1}(\mathbb{R})$, we apply the Cauchy-Lipschitz-Picard Theorem [4] to obtain that for any initial data $(\overline{M}, \overline{N}) \in L^2(0, 1) \times L^2(0, 1)$ the problem (1.9) admits a unique solution $\sigma \in H^2(0, \infty; L^2(0, 1))$. We define the Eulerian flow

$$\Psi(t, M, N) = (\Psi^1(t, M, N), \Psi^2(t, M, N)) = (\sigma_t, \dot{\sigma}_t). \qquad (1.11)$$

We can then easily check that

$$f_t := (M(t, \cdot), \dot{M}(t, \cdot))_\# \chi_{(0,1)} \text{ with } f_0 = (M_0, N_0)_\# \chi_{(0,1)}$$

satisfies (1.8). Note that (1.9) is Hamiltonian and the energy $E(t) := H(\Psi(t, M, N))$ is conserved: $E(0) = E(t)$. Here, the Hamiltonian and the Lagrangian $H, L : L^2(0, 1) \times L^2(0, 1) \to \mathbb{R}$ are given by

$$H(M, N) = \frac{1}{2}\|N\|^2 + \frac{1}{2}\mathcal{W}(M), \quad L(M, N) = \frac{1}{2}\|N\|_{v_0}^2 - \frac{1}{2}\mathcal{W}(M) \quad (1.12)$$

where

$$\mathcal{W}(M) := \int_{(0,1)^2} W(Mz - Mw)\mathrm{d}z\mathrm{d}w.$$

It is not a loss of generality to assume that $W(0) = 0$ and W is even. Indeed, we may substitute W by $W - W(0)$ without altering (1.9). Also, substituting W by $z \to [W(z) + W(-z)]/2$ will not alter $\mathcal{W}$. In order to make some computations simpler, we further assume that

$$W(z) = W(-z) \leq W(0) = 0 \text{ for all } z \in \mathbb{R}. \tag{1.13}$$

The only restrictive assumption here is that the maximum of W is attained at 0.

Let $\mathcal{G}$ be the set of bijections $G : [0, 1] \to [0, 1]$ such that G, G^{-1} are Borel and push ν_0 forward to itself. The group $\mathcal{G}$ acts on $L^2(0, 1)$: $(G, M) \in \mathcal{G} \times L^2(I) \to M \circ G$. It also acts on the topological subspace $L^2_{\mathbb{Z}}(0, 1)$ and so, induces a natural action on $\mathbb{T}$ and on the tangent bundle $L^2(0, 1) \times L^2(0, 1)$. Note that L and H are invariant under the action of $\mathcal{G}$.

Our goal is to prove the following result: for each fixed positive integer n, let $\mathcal{C}_n$ be the set of real valued functions M on $[0, 1]$, constant on each subinterval $I_i^n := ((i - 1)/n, i/n), i = 1, \cdots, n$. Let $L^n \in C^2(\mathbb{R}^n \times \mathbb{R}^n)$ and $H^n \in C^2(\mathbb{R}^n \times \mathbb{R}^n)$ be the Lagrangian and Hamiltonian defined in (3.1) and (3.2). We fix $c \in \mathbb{R}$. The standard Hamilton-Jacobi theory provides us with an explicit way for constructing $u^n(\cdot; c) \in C(\mathbb{R}^n)$, $\mathbb{Z}^n$-periodic viscosity solutions of $H^n(x, \nabla u^n(x; c) + \mathbf{c}_n) = c^2/2$ in $\mathbb{R}^n$, where $\mathbf{c}_n := (c, c, \ldots, c) \in \mathbb{R}^n$. Let us introduce the notation

$$L_c(M, N) := L(M, N) - c \int_0^1 M\mathrm{d}x. \tag{1.14}$$

Theorem 1.1. *There exists a rearrangement invariant $\mathcal{U}(\cdot; c)$, Lipschitz continuous in the strong $L^2(0, 1)$-topology satisfying the following properties: for all $n \geq 1$ integer and $x \in \mathbb{R}^n$, $u^n(x; c) = \mathcal{U}\left(\sum_{i=1}^n x_i \chi_{I_i^n}; c\right)$. Furthermore, $\mathcal{U}(\cdot; c)$ is a viscosity solution for*

$$H(M, \nabla_{L^2}\mathcal{U}(M; c) + c) = \frac{c^2}{2} \tag{1.15}$$

and $\mathcal{U}(M; c) \in C(\mathbb{T})$. Similarly, there exists a rearrangement invariant $\mathcal{U}_*(\cdot; c)$ Lipschitz continuous in the strong $L^2(0, 1)$-topology which satisfies the following conditions: for each nondecreasing $M \in L^2(0, 1)$ there exists a so-called calibrated curve σ^c associated to $\mathcal{U}_*(\cdot; c)$ in the sense *that $\sigma^c \in H^2(0, \infty; L^2(0, 1))$, $\sigma_0^c = M$ and whenever $T > 0$,*

$$\mathcal{U}_*(\sigma_T^c; c) = \int_0^T L_c(\sigma_t^c, \dot{\sigma}_t^c)dt + \mathcal{U}_*(M; c) + \frac{1}{2}c^2 T.$$

Furthermore, for all $\sigma \in H^2(0, \infty; L^2(0, 1))$ we have

$$\mathcal{U}_*(\sigma_T; c) \leq \int_0^T L_c(\sigma_t, \dot{\sigma}_t)dt + \mathcal{U}_*(\sigma_0; c) + \frac{1}{2}c^2 T.$$

It is proven in [16] that the following corollaries are direct consequences of Theorem 1.1.

Corollary 1.2. *For each $c \in \mathbb{R}$ and each $M \in L^2(0, 1)$ which is monotone nondecreasing, there exists $N \in L^2(I)$ such that*

$$\sup_{t>0} \sqrt{t} \left\| \frac{\Psi^1(t, M, N)}{t} - c \right\|_{v_0} < \infty, \qquad \lim_{t \to \infty} \Psi^2(t, M, N) = c.$$

Corollary 1.3. *Given $c \in \mathbb{R}$ and a Borel probability measure μ on $\mathbb{R}$ of bounded second moment, there exists a path $t \to \rho_t \in AC_{loc}^2(0, \infty; \mathcal{P}_2(\mathbb{R}))$ and $u : (0, \infty) \times \mathbb{R} \to \mathbb{R}$ Borel satisfying the following properties: $u_t \in L^2(\rho_t)$ for $\mathcal{L}^1$-almost every $t > 0$, and $\rho_0 = \mu$. The pair (ρ, u) satisfies the Euler system*

$$\begin{cases} \partial_t(\rho_t u_t) + \partial_x(\rho_t u_t^2) = -\rho_t \partial_x P_t \\ \partial_t \rho_t + \partial_x(\rho_t u_t) = 0 \\ P_t(x) = \int_{\mathbb{R}} W(x - y)d\rho_t(y). \end{cases} \tag{1.16}$$

Furthermore,

$$\sup_{t>0} \sqrt{t}\|\mathbf{id}/t - c\|_{\rho_t} < \infty, \qquad \lim_{t \to \infty} \|u_t - c\|_{\rho_t} = 0. \tag{1.17}$$

We have chosen the Vlasov system as a simple model to illustrate the use of the weak KAM theory for understanding qualitative behavior of PDEs appearing in kinetic theory, for several reasons. Firstly, they provide a simple link between finite and infinite dimensional systems. Secondly, they are one of the most frequently used kinetic models in statistical mechanics. Existence and uniqueness of global solutions for the initial value problem are well understood [3,9,19]. In this paper we have searched for special solutions which allow for a connection with a more conventional way of regarding (1.8) as Hamiltonian. We assume the initial data to be of the form $f_0 = (M, N)_{\#}\nu_0$ where $M, N \in L^2(X)$ so that the unique solution of (1.8) retains the same structure.

ACKNOWLEDGEMENTS. We are indebted to Luigi De Pascale who took notes during the lectures and typed part of this manuscript. We would like to thank the organizers, L. Ambrosio, G. Buttazzo, N. Fusco and G. Savaré for giving us the opportunity to present these new results and making arrangements for their publication.

2 Effective Hamiltonian

In this section we define the effective Hamiltonian $\overline{H}$ in our infinite-dimensional setting and compute $\overline{H}(c)$. The choice of constant functions as "rotation numbers" in this context is fully justified in [16].

We begin by recalling some results from [17], adapted to our setting. In [17] we proved the existence of an infinite-dimensional *effective Lagrangian* under the following assumption: suppose L is a Lagrangian on $L^2(0, 1) \times L^2(0, 1)$ satisfying the growth conditions

$$c\|N\|^2 \leq L(M, N) \leq C\big(1 + \|N\|^2\big), \quad \text{for all } (M, N) \in [L^2(X)]^2, \quad (2.1)$$

where c, C are given positive constants. Assume L is $L^2_{\mathbb{Z}}(0, 1)$-periodic in M, i.e.

$$L(M+Z, N) = L(M, N) \text{ for all } Z \in L^2_{\mathbb{Z}}(0, 1), M, N \in L^2(0, 1). \quad (2.2)$$

Assume further that there exists $\Lambda > 0$ such that

$$L(M, N_1) - L(M, N_2) \leq \Lambda \int_{\{N_1 \neq N_2\}} |N_1|^2 dx \qquad (2.3)$$

for all $M, N_1, N_2 \in L^2(0, 1)$. Also, it satisfies, for some continuous, nondecreasing $\omega : \mathbb{R} \to \mathbb{R}$ such that $\omega(0) = 0$,

$$|L(M_1, N) - L(M_2, N)| \leq \omega\big(\mathcal{L}^1(\{M_1 \neq M_2\})\big) \qquad (2.4)$$

for all $M_1, M_2, N \in L^2(0, 1)$. Fix $T > 0$ and consider

$$\mathcal{H} := H^1(0, T; L^2(0, 1))$$

endowed with the topology τ given by

$$\begin{aligned} M_n \xrightarrow{\tau} M \iff &\|M_n - M\|_{L^2((0,T) \times \Omega)} \\ &\to 0 \text{ and } \{\dot{M}_n\} \text{ is bounded in } L^2((0, T) \times X) \end{aligned} \qquad (2.5)$$

for every $\Omega \subset\subset X$. By following mostly the techniques in [10], we have proved in [17] (in even more generality) that

$$\int_0^T \overline{L}(\dot{\sigma})dt = \Gamma(\tau) \lim_{\varepsilon \to 0} \int_0^T L\left(\frac{\sigma}{\varepsilon}, \dot{\sigma}\right)dt$$

for

$$\overline{L}(N) := \liminf_{T \to \infty} \inf_{\phi \in \mathcal{H}_0} \int_0^T L(tN + \phi(t), N + \dot{\phi}(t))dt, \qquad (2.6)$$

where $\Gamma(\tau)$ denotes the Γ-convergence with respect to the topology τ. The set $\mathcal{H}_0$ represents all functions in $\mathcal{H}$ with null trace. The continuity of the map $\overline{L}$ with respect to the strong $L^2(0, 1)$ topology was obtained as a consequence of its convexity and local boundedness.

Definition 2.1. The map $\overline{L}$ is called the effective Lagrangian corresponding to L. Its Legendre transform defined for $\zeta \in L^2(0, 1)$ by

$$\overline{H}(\zeta) = \sup_{\xi \in L^2(0,1)} \left\{ \langle \xi, \zeta \rangle_{L^2(X)} - \overline{L}(\xi) \right\}$$

is called the effective Hamiltonian associated to H (the Legendre transform of L).

We proved in [17] that the viscosity solutions (given by the Lax-Oleinik variational formula) for the evolutionary Hamilton-Jacobi equations with oscillating Hamiltonians $H(\cdot/\epsilon, \cdot)$ converge to the Hopf-Lax solution of the Hamilton-Jacobi equation with the effective $\overline{H}$. In the classical but unpublished paper [18] the authors arrived to the effective Hamiltonian by performing this homogenization. They showed that for every $P \in \mathbb{R}^n$ (rotation vector) there exists a unique $\lambda \in \mathbb{R}$ such that the *cell problem*

$$H(x, \nabla u(x) + P) = \lambda$$

admits a periodic viscosity solution; then $\overline{H}$ was defined by $\lambda = \overline{H}(P)$. E [10] showed that if $H(x, p)$ is convex in p, then one can use Lax-Oleinik's representation for the viscosity solutions to obtain the homogenization result in [18]. Indeed, in his approach, $\overline{L}$ is obtained first as an object giving the Γ-limit of oscillating integral functionals. Note that in [17] we followed E's approach and arrived to the effective Hamiltonian by means of the Legendre transform of the effective Lagrangian.

We now return to $L : L^2(0, 1) \times L^2(0, 1) \to \mathbb{R}$ given by (1.12). For $c \in \mathbb{R}$, not only are we able to compute $\overline{H}(c)$ explicitly, but we will see in the next section that its discrete counterparts have precisely the same value. This feature turns out to be crucial for our analysis, as our approach is of the finite-to-infinite-dimensions kind.

Proposition 2.2. *If we identify $c \in \mathbb{R}$ with the constant function $f \equiv c$ over $(0, 1)$, then*

$$\overline{L}(c) = \overline{H}(c) = \frac{1}{2}c^2 \text{ for all } c \in \mathbb{R}. \tag{2.7}$$

Proof. Since $W \leq 0$, (1.12) implies $L(M, N) \geq \|N\|^2/2$ for all $M, N \in L^2(0, 1)$. From (2.6) we deduce $\overline{L}(N) \geq \|N\|^2/2$, so

$$c \int_0^1 N \, dx - \overline{L}(N) \leq c \int_0^1 N \, dx - \frac{1}{2}\int_0^1 N^2 \, dx \leq \frac{1}{2}c^2 \text{ for all } N \in L^2(0, 1).$$

Thus, $\overline{H}(c) \leq c^2/2$ and we now need to prove the opposite inequality. For this we observe that

$$\inf_{\phi \in \mathcal{H}} \int_0^T L(tc + \phi(t), c + \dot\phi(t)) dt = \int_0^T L(tc, c) dt = \frac{1}{2}c^2$$

for all $T > 0$ because $W \leq 0 = W(0)$ and the infimum is taken over $H^1(0, T; L^2(0, 1))$ functions such that $\phi(0) = \phi(T) = 0$. According to (2.6), we obtain $\overline{L}(c) = c^2/2$. Therefore, $\overline{L}(c) + \overline{H}(c) \geq c^2$ yields $\overline{H}(c) \geq c^2/2$ which concludes the proof. $\qquad\square$

3   From finite to infinite-dimensions

In this section we introduce the discrete versions of the particle interaction Lagrangian and Hamiltonian. We study the corresponding cell problems, then we show that the viscosity solutions obtained by a linear perturbation approximation argument are finite-dimensional restrictions of a rearrangement invariant, periodic, Lipschitzian functional on $L^2(0, 1)$.

3.1   Discrete Hamiltonians

We endow $\mathbb{R}^n$ with the inner product $\langle x, y \rangle_n = x \cdot y/n$, denote by $|\cdot|_n$ the induced norm and by ∇_n the induced gradient. Let us define $L^n : \mathbb{R}^n \times \mathbb{R}^n \to \mathbb{R}$ by

$$L^n(x, v) = \frac{1}{2}|v|_n^2 - \frac{1}{2n^2}\sum_{i,j=1}^n W(x_i - x_j). \tag{3.1}$$

Its Legendre transform is, clearly, $H^n : \mathbb{R}^n \times \mathbb{R}^n \to \mathbb{R}$ defined by

$$H^n(x, \mathbf{p}) = \frac{1}{2}|\mathbf{p}|_n^2 + \frac{1}{2n^2}\sum_{i,j=1}^n W(x_i - x_j). \tag{3.2}$$

Note that $L^n(x, v) = L(M_n, N_n)$, where M_n and N_n are piecewise constant $M_n \equiv x_i$, $N_n \equiv v_i$ on the n-regular partitions of X. One can easily adapt the proof of Proposition 2.2 above to prove:

Lemma 3.1. *If we denote by* $c^n := (c, \ldots, c) \in \mathbb{R}^n$

$$\overline{H^n}(\mathbf{c}^n) = \frac{1}{2}c^2 \text{ for all integers } n \geq 1 \text{ and all } c \in \mathbb{R}. \qquad (3.3)$$

We know from the classical, finite-dimensional theory, that (3.3) implies that $c^2/2$ is the unique real number λ for which the equation

$$H^n(x, \nabla_n v(x) + c^n) = \lambda \qquad (3.4)$$

admits a $\mathbb{Z}^n$-periodic viscosity solution [11,12] denoted by $u^n(\cdot; c)$. Since these solutions are, in general, not unique, we choose specific ones, obtained by a standard approximation argument.

Remark 3.2. For a generic Lagrangian l and its associated Hamiltonian h, one can obtain [5,13] a periodic viscosity solution for the cell problem (1.7) the following way: for $\mathbf{c} \in \mathbb{R}^n$ define $\tilde{h}(x, \mathbf{p}) = h(x, \mathbf{c} - \mathbf{p})$, so that its Legendre transform is $\tilde{l}(x, \mathbf{v}) = \mathbf{c} \cdot \mathbf{v} + l(x, -\mathbf{v})$. Then, for $\epsilon > 0$ there exists a unique periodic viscosity solution for $\epsilon w_\epsilon + \tilde{h}(x, -\nabla w_\epsilon) = 0$ such that the pair $(w_\epsilon - \min w_\epsilon, -\epsilon w_\epsilon)$ converges as $\epsilon \downarrow 0$ (possibly, up to subsequence) to $(w, \tilde{h}(\mathbf{c}))$ uniformly on $\mathbb{R}^n$, where w is a periodic solution for the cell problem. It is known [13] that w_ϵ admits the representation

$$w_\epsilon(x) = \inf_{\sigma(0)=x} \int_0^\infty e^{-\epsilon s} \tilde{l}(\sigma(s), \dot{\sigma}(s)) ds.$$

This fact will be used below.

We now return to the L^n case and introduce the Lagrangian L_c^n along with its corresponding Hamiltonian H_c^n by

$$L_c^n(x, \mathbf{v}) = L^n(x, \mathbf{v}) - \langle c^n, \mathbf{v} \rangle_n, \quad H_c^n(x, \mathbf{p}) = H^n(x, c^n + \mathbf{p}). \qquad (3.5)$$

Note that they are, indeed, Legendre conjugates. Then (3.4) becomes

$$H_c^n(x, \nabla_n v(x)) = \overline{H^n}(\mathbf{c}^n) = \frac{1}{2}c^2. \qquad (3.6)$$

For $\epsilon > 0$ one looks at the unique viscosity solution $u_\epsilon^n(\cdot; c)$ for

$$\epsilon v + H_c^n(x, \nabla_n v(x)) = 0 \qquad (3.7)$$

and shows [18] that $u_\epsilon^n(\cdot; c) - \min u_\epsilon^n(\cdot; c)$ converges uniformly to some $u^n(\cdot; c)$ which is a viscosity solution for (3.6) and, thus, for (3.4). By Remark 3.2, these unique $u_\epsilon^n(\cdot; c)$ have the following representation formula

$$u_\epsilon^n(x; c) = \inf_{\sigma(0)=x} \int_0^\infty e^{-\epsilon s} \tilde{L}_c^n(\sigma(s), \dot{\sigma}(s)) ds \qquad (3.8)$$

where $\sigma \in H^1(0, \infty; \mathbb{R}^n)$ and $\tilde{L}_c^n(x, \mathbf{v}) = L^n(x, \mathbf{v}) + \langle c^n, \mathbf{v} \rangle_n$. Due to the permutation invariance of $\tilde{L}_c^n$, we infer that

$$u_\epsilon^n(x; c) = u_\epsilon^n(x_\tau; c) \text{ for any permutation of } n \text{ letters } \tau, \qquad (3.9)$$

where we recall that $x^\tau = (x_{\tau(1)}, \ldots, x_{\tau(n)})$. Let us now prove a useful lemma.

Lemma 3.3. *For every $c \in \mathbb{R}$, $\epsilon > 0$ and every positive integer n,*

$$\min_{x \in \mathbb{R}^n} u_\epsilon^n(x; c) = u_\epsilon^n(0; c) = -\frac{c^2}{2\epsilon}.$$

Proof. According to (3.8),

$$u_\epsilon^n(x; c) = \inf_{\sigma(0)=x} \int_0^\infty e^{-\epsilon s} \left[\frac{1}{2} |\dot{\sigma}|_n^2 + \langle \dot{\sigma}, c^n \rangle_n - \frac{1}{2n^2} \sum_{i,j=1}^n W(\sigma_i - \sigma_j) \right] ds$$

$$= \inf_{\sigma(0)=x} \int_0^\infty e^{-\epsilon s} \left[-\frac{c^2}{2} + \frac{1}{2n} \sum_{i=1}^n |\dot{\sigma}_i + c|^2 - \frac{1}{2n^2} \sum_{i,j=1}^n W(\sigma_i - \sigma_j) \right] ds.$$

Since $W \le 0$, clearly the minimum with respect to x is attained at $x = 0$ when $\sigma_i(s) = -cs$ for all $s \ge 0$. Thus, the conclusion follows.   □

Next, we prove a "consistency" result.

Lemma 3.4. *Let m, n be positive integers. Then*

$$u_\epsilon^{mn}(x_1^m, \ldots, x_n^m; c) = u_\epsilon^n(x; c), \qquad (3.10)$$

where $x_j^m := (x_j, \ldots, x_j) \in \mathbb{R}^m$ for $1 \le j \le n$.

Proof. Without loss of generality, we may assume $c = 0$ and drop the dependence on c in the notation. We have

$$u_\epsilon^{mn}(x)$$

$$= \inf_{\sigma_{mn}} \int_0^\infty e^{-\epsilon s} \left[\frac{1}{2mn} \sum_{i=1}^{mn} |\dot{\sigma}_{mn}^i|^2 - \frac{1}{2m^2n^2} \sum_{i,j=1}^{mn} W(\sigma_{mn}^i - \sigma_{mn}^j) \right] ds, \qquad (3.11)$$

where $\sigma_{mn}^i(0) = x_j$ for all $1 \le j \le n$ and all $(j-1)m+1 \le i \le jm$. Now let us consider the expression on the right hand side under the restriction

$\sigma_{mn}^{i_1} = \sigma_{mn}^{i_2} =: \sigma_n^j$ for any $1 \leq j \leq n$ and every $(j-1)m + 1 \leq i_1, i_2 \leq jm$. Then,

$$\frac{1}{2mn} \sum_{i=1}^{mn} |\dot\sigma_{mn}^i|^2 - \frac{1}{2m^2n^2} \sum_{i,j=1}^{mn} W(\sigma_{mn}^i - \sigma_{mn}^j) = \frac{1}{2n} \sum_{j=1}^{n} |\dot\sigma_n^j|^2$$
$$- \frac{1}{2n^2} \sum_{i,j=1}^{n} W(\sigma_n^i - \sigma_n^j),$$

which means that if we take the infimum in (3.11) with respect to x under this restriction we get

$$u_\epsilon^{mn}(x_1^m, \ldots, x_n^m) \leq u_\epsilon^n(x).$$

To prove the opposite inequality, denote by $J_k = \{(k-1)m + 1, \ldots, km\}$ for $1 \leq k \leq n$. Since $W \leq 0$, after throwing away some nonnegative terms, one has

$$\int_0^\infty e^{-\epsilon s} \left[\frac{1}{2mn} \sum_{i=1}^{mn} |\dot\sigma_{mn}^i|^2 - \frac{1}{2m^2n^2} \sum_{i,j=1}^{mn} W(\sigma_{mn}^i - \sigma_{mn}^j) \right] ds$$

$$\geq m^{-n} \int_0^\infty e^{-\epsilon s} \sum_{\mathbf{k} \in J_1 \times \ldots \times J_n} \left[\frac{1}{2n} \sum_{i=1}^{n} |\dot\sigma_{mn}^{k_i}|^2 - \frac{1}{2n^2} \sum_{i,j=1}^{n} W(\sigma_{mn}^{k_i} - \sigma_{mn}^{k_j}) \right] ds$$

$$\geq m^{-n} \sum_{\substack{\mathbf{k} \in J_1 \times \ldots \times J_n \\ \sigma_{mn}^{k_i}(0) = x_i, \\ 1 \leq i \leq n}} \inf \int_0^\infty e^{-\epsilon s} \left[\frac{1}{2n} \sum_{i=1}^{n} |\dot\sigma_{mn}^{k_i}|^2 - \frac{1}{2n^2} \sum_{i,j=1}^{n} W(\sigma_{mn}^{k_i} - \sigma_{mn}^{k_j}) \right] ds$$

$$= m^{-n} m^n u_\epsilon^n(x) = u_\epsilon^n(x).$$

We conclude by taking the infimum in the left hand side. $\square$

By Lemma 3.3 we infer

$$u_\epsilon^n(\cdot; c) + \frac{c^2}{2\epsilon} \to u^n(\cdot; c) \quad \text{uniformly in } \mathbb{R}^n \text{ as } \epsilon \downarrow 0.$$

Due to the periodicity of L_c^n and the uniqueness of $u_\epsilon^n(\cdot; c)$, it follows that $u_\epsilon^n(\cdot; c)$ is periodic. Lemma 3.3 and (3.9) imply that $u^n(\cdot; c)$ is also $\mathbb{Z}^n$-periodic and permutation invariant. Furthermore, Lemma 3.4 implies

$$u^{mn}(x_1^m, \ldots, x_n^m; c) = u^n(x; c) \quad \text{for all positive integers } m, n. \quad (3.12)$$

It is also known that $u^n(\cdot; c)$ is Lipschitz on $\mathbb{R}^n$, so it is differentiable a.e. and the equation (3.4) is satisfied pointwise at the points of differentiability. One can easily see then that the Lipschitz constant κ_n satisfies $0 < \kappa_n \leq \sqrt{c^2 - 2 \inf V} =: \kappa$.

3.2 An infinite-dimensional extension

Let us now consider partitioning the interval $X = (0, 1)$ into n equal subintervals and denote by $\mathcal{C}_n$ the set of all real-valued functions defined on $(0, 1)$ and constant on each such subinterval. Any function $f \in \mathcal{C}_n$ can be identified with a vector $x_f^n \in \mathbb{R}^n$ by its values. We now define

$$\tilde{U} : \cup_{n \geq 1} \mathcal{C}_n =: \mathcal{C} \to \mathbb{R} \text{ by } \tilde{U}(f; c)$$
$$= u^{n_0}(x_f^{n_0}; c) \text{ whenever } f \in \mathcal{C}_n. \tag{3.13}$$

According to Lemma 3.4, not only is this functional well-defined, but it is also Lipschitz on $\mathcal{C}$ with respect to the strong $L^2(0, 1)$-norm. Indeed, let $f, g \in \mathcal{C}$. Then $f \in \mathcal{C}_n$ and $g \in \mathcal{C}_m$ for some positive integers m, n, so $f, g \in \mathcal{C}_{mn}$. Thus,

$$|\tilde{U}(f; c) - \tilde{U}(g; c)| = |u^{mn}(x^{mn}; c) - u^{mn}(\mathbf{y}^{mn}; c)| \leq \kappa |x^{mn} - \mathbf{y}^{mn}|_{mn}.$$

But $|x^{mn} - \mathbf{y}^{mn}|_{mn} = \|f - g\|_{L^2(0,1)}$, so the claim is proved. Due to the density of $\mathcal{C}$ in $L^2(0, 1)$, we conclude that $\tilde{U}$ can be uniquely extended by continuity to $L^2(0, 1)$. More precisely:

Proposition 3.5. *For any $c \in \mathbb{R}$ there exists a unique $U(\cdot; c): L^2(0, 1) \to \mathbb{R}$ which is Lipschitz continuous with $\mathrm{Lip}(U(\cdot; c)) \leq \kappa$ and such that $U(\cdot; c)|_{\mathcal{C}_n} = u^n(\cdot; c)$.*

In the next section we will prove that U is the viscosity solution we are looking for. Before that, let us show that it has some "nice" properties inherited from u^n.

Proposition 3.6. *For any $c \in \mathbb{R}$ the functional $U(\cdot; c)$ is RI and $L_{\mathbb{Z}}^2(0, 1)$-periodic.*

Proof. We give up the c-dependence to unburden notation. Let $M \in L^2(0, 1)$ and $\tilde{M}$ be its monotone rearrangement. Take a sequence of maps $M_n \in \mathcal{C}_n$ converging to M in $L^2(0, 1)$, denote by $\tilde{M}_n$ their monotone rearrangements and set $\mu_n = M_{n\#}\nu_0$, $\mu = M_\#\nu_0$. By one-dimensional optimal transport [16], we have

$$\|\tilde{M}_n - \tilde{M}\| = W_2(\mu_n, \mu) \leq \|M_n - M\|$$

which gives $\tilde{M}_n \to \tilde{M}$ in $L^2(0, 1)$. Here, W_2 denotes the 2-Wasserstein distance on the real line (*cf.* [2, 15, 27]). Since u^n is permutation invariant, we conclude $U(M_n) = U(\tilde{M}_n)$ which, due to the continuity of U, implies $U(M) = U(\tilde{M})$. Thus, U is rearrangement invariant. To prove periodicity take $Z \in L_{\mathbb{Z}}^2(0, 1)$ and a sequence $Z_n \in L_{\mathbb{Z}}^2(0, 1)$ piecewise constant on the n-equipartition of $(0, 1)$ such that $Z_n \to Z$ in $L^2(0, 1)$. Then $M_n + Z_n \to M + Z$ in $L^2(0, 1)$, so $U(M_n + Z_n) \to U(M + Z)$. But the $\mathbb{Z}^n$-periodicity of u^n yields $U(M_n + Z_n) = U(M_n)$ and the continuity of U concludes the proof. □

Remark 3.7. In the proof we have used the fact that any $Z \in L_{\mathbb{Z}}^2(0, 1)$ is the L^2-limit of functions that are integer-valued, piecewise constant on the n-equipartition of $(0, 1)$. Indeed, to see that, note that we may first approximate Z by functions taking on only finitely many values. So it suffices to prove the statement for indicator functions of Borel sets $A \subset 0, 1$. Since the Lebesgue measure is Borel regular, it is enough to consider open sets $O \subset (0, 1)$. Furthermore, one can reduce these open sets to finite unions of disjoint open subintervals of $(0, 1)$. For such sets, the property is easy to prove.

4   The weak KAM theorem

Here we shall prove Theorem 1.1, *i.e.* we shall show that $U(\cdot; c)$ constructed in the previous section provides a viscosity solution for (1.15).

Definition 4.1. Let V be a real valued proper functional defined on $L^2(0, 1)$ with values in $\mathbb{R} \cup \{\pm\infty\}$. Let $M_0 \in L^2(0, 1)$ and $\xi \in L^2(0, 1)$. We say that ξ belongs to the (Fréchet) subdifferential of V at M_0 and we write $\xi \in \partial.V(M_0)$ if

$$V(M) - V(M_0) \geq \langle \xi, M - M_0 \rangle + o(\|M - M_0\|)$$

for all $M \in L(0, 1)$.

   We say that ξ belongs to the superdifferential of V at M_0 and we write $\xi \in \partial^. V(M_0)$ if $-\xi \in \partial.(-V)(M_0)$.

Remark 4.2. As expected, when the sets $\partial.V(M_0)$ and $\partial^. V(M_0)$ are both nonempty, then they coincide and consist of a single element. That element is the L^2-gradient of V at M_0, denoted by $\nabla_{L^2} V(M_0)$.

4.1   Viscosity solutions; solution semigroup

We can now define [6] the notion of viscosity solution for a general Hamilton-Jacobi equation of the type

$$F(M, \nabla_{L^2} U(M)) = 0. \tag{HJ}$$

Definition 4.3. Let $V : L^2(0, 1) \to \mathbb{R}$ be continuous.

(i) We say that V is a viscosity subsolution for (HJ) if

$$F(M, \zeta) \leq 0 \text{ for all } M \in L^2(0, 1) \text{ and all } \zeta \in \partial^. V(M). \quad (4.1)$$

(ii) We say that V is a viscosity supersolution for (HJ) if

$$F(M, \zeta) \geq 0 \text{ for all } M \in L^2(0, 1) \text{ and all } \zeta \in \partial. V(M). \quad (4.2)$$

(iii) We say that V is a viscosity solution for (HJ) if V is both a subsolution and a supersolution for (HJ).

Remark 4.4. If U is a viscosity solution, then, in view of Remark 4.2, we deduce that (HJ) is satisfied at all $M \in L^2(0, 1)$ where $\partial. U(M) \cap \partial^. U(M) \neq \emptyset$, which are precisely the points where U is differentiable.

Let $M \in L^2(0, 1)$ and $V : L^2(0, 1) \to \mathbb{R}$ be continuous and bounded. For $t \geq 0$ define the operator $T_{L,t}$ on the space of uniformly continuous and bounded functionals $BUC(L^2(0, 1))$ by

$$T_{L,t}V(M)$$
$$:= \inf_{S(t)=M} \left\{ V(S(0)) + \int_0^t L(S(\tau), \dot{S}(\tau)) d\tau : S \in H^1(0, t; L^2(0, 1)) \right\}. \quad (4.3)$$

Observe that $t \to T_{L,t}$ defines a (backward) semigroup on $[0, \infty)$. Furthermore, $\mathcal{U}(t, M) =: T_{L,t}V(M)$ yields the unique viscosity solution [6,7] for the Cauchy problem associated with the evolutionary Hamilton-Jacobi equation

$$\partial_t \mathcal{U}(t, M) + H(M, \nabla_{L^2}\mathcal{U}(t, M)) = 0, \quad \mathcal{U}(0, M) = V(M).$$

As a consequence, we have the following:

Proposition 4.5. *The map $\mathcal{V} \in BUC$ is a fixed point for $\{T_{L,t}\}_{t \geq 0}$, i.e.*

$$T_{L,t}\mathcal{V} = \mathcal{V} \text{ for all } t \geq 0 \quad (4.4)$$

if and only if $\mathcal{V}$ is a viscosity solution for $H(M, \nabla_{L^2}\mathcal{V}(M)) = 0$.

Indeed, if we put $V(t, M) := \mathcal{V}(M)$, then according to the above discussion V solves (in the viscosity sense) the Cauchy problem with initial data $\mathcal{V}$. Since this V is, in fact, time-independent, we deduce that it is a viscosity solution for the stationary HJ equation. Similarly, we obtain that U constructed at the end of the previous section satisfies the requirements of Theorem 1.1 if it has the following property.

Proposition 4.6. *For any $c \in \mathbb{R}$ let $U(\cdot; c)$ be the functional from Proposition 3.5. Then*

$$T_{L_c,t}U(\cdot; c) = U(\cdot; c) - \frac{1}{2}c^2 t \text{ for all } t \geq 0, \tag{4.5}$$

where $I_{\cdot c}$ is defined in (1.14).

The goal of the remainder of this section is proving Proposition 4.6. To achieve this, we consider the discrete L^n and use it to define (we use $T_{c,t}^n$ instead of $T_{L_c^n,t}^n$ to unburden notation)

$$T_{c,t}^n v(x)$$
$$:= \inf_{\sigma(t)-x} \left\{ v(\sigma(0)) + \int_0^t L_c^n(\sigma(\tau), \dot{\sigma}(\tau)) d\tau : \sigma \in H^1(0, t; \mathbb{R}^n) \right\}, \tag{4.6}$$

where L_c^n is defined in (3.5). Likewise, any viscosity solution of $H^n(x, \nabla_n v(x) + c^n) = c^2/2$ satisfies the n-dimensional version of (4.5). We deduce that, in particular, the n-dimensional approximations (restrictions, rather) u^n of U satisfy

$$T_{c,t}^n u^n(\cdot; c) = u^n(\cdot; c) - \frac{1}{2}c^2 t \text{ for all } t \geq 0. \tag{4.7}$$

We would like to use this to prove (4.5) by passing to the limit as $n \to \infty$ in some sense.

Remark 4.7. Note that if we further simplify notation by setting $T_t := T_{L_0,t}$ and $T_t^n := T_{0,t}^n$, easy calculations show that (4.7) becomes

$$T_t^n \bar{u}^n(\cdot; c) = \bar{u}^n(\cdot; c) \text{ for all } t \geq 0, \text{ where } \bar{u}^n(\cdot; c) := u^n(\cdot; c) + \langle \cdot, c^n \rangle_n.$$

Likewise, (4.5) becomes

$$T_t \overline{U}(\cdot; c) = \overline{U}(\cdot; c) \text{ for all } t \geq 0,$$

$$\text{where } \overline{U}(M; c) := U(M; c) + c \int_0^1 M dx.$$

4.2 The main result

Again, in this subsection, we consider the case $c = 0$ without loss of generality. Thus, we can use the notation from Remark 4.7.

Lemma 4.8. *Let $U : L^2(0, 1) \to \mathbb{R}$ be Lipschitz continuous. Then, for any $t > 0$, $T_t U$ is uniformly continuous on $L^2(0, 1)$.*

Proof. Let $\varepsilon > 0$ be fixed. Take $\delta > 0$ (to be fixed later), M_1, $M_2 \in L^2(0, 1)$ such that $\|M_1 - M_2\| \leq \delta$. By definition, there exists $S_1 \in H^1(0, t; L^2(0, 1))$ with $S_1(t) = M_1$ such that

$$U(S_1(0)) + \int_0^t L(S_1(s), \dot{S}_1(s))ds \leq T_t U(M_1) + \delta. \qquad (4.8)$$

Define

$$S^\delta(s) = \begin{cases} S_1(s) & \text{if } 0 \leq s \leq t - \delta \\ \dfrac{s - t + \delta}{\delta}(M_2 - M_1) + S_1(s), & \text{if } t - \delta \leq s \leq t, \end{cases}$$

a path connecting $S_1(0)$ and M_2. Thus,

$$T_t U(M_2) \leq U(S_1(0)) + \int_0^t L(S_1, \dot{S}_1)ds - \int_{t-\delta}^t L(S_1, \dot{S}_1)ds$$

$$+ \int_{t-\delta}^t L(S^\delta, \dot{S}^\delta)ds$$

$$\leq T_t U(M_1) + \delta + C\delta + \frac{1}{2}\int_{t-\delta}^t (\|\dot{S}^\delta\|^2 - \|\dot{S}_1\|^2)ds$$

$$\leq T_t U(M_1) + C\delta + \frac{1}{2\delta}\|M_2 - M_1\|^2 + \frac{\|M_2 - M_1\|}{\delta}\int_{t-\delta}^t \|\dot{S}_1(s)\|ds$$

$$\leq T_t U(M_1) + C\delta + \frac{1}{2\delta}\|M_2 - M_1\|^2 + \frac{\|M_2 - M_1\|}{\sqrt{\delta}}\left(\int_0^t \|\dot{S}_1(s)\|^2 ds\right)^{1/2}.$$

But if we consider the constant path M_1 in the variational principle we get

$$T_t U(M_1) \leq U(M_1) - \frac{t}{2}\iint_{X^2} V(M_1(x) - M_1(y))dydx,$$

so, in view of (4.8), we obtain

$$\frac{1}{2}\int_0^t \|\dot{S}_1(s)\|^2 ds \leq \delta + Ct + U(M_1) - U(S_1(0)).$$

Now we use

$$\|M_1 - S_1(0)\| \leq \int_0^t \|\dot{S}_1(s)\|ds \leq \sqrt{t}\left(\int_0^t \|\dot{S}_1(s)\|^2 ds\right)^{1/2}$$

and the fact that U is Lipschitz to infer that $\int_0^t \|\dot{S}_1(s)\|^2 ds$ is bounded by some $C(\delta, t)$ (increasing in each variable). So, for $\delta < 1$, $\int_0^t \|\dot{S}_1(s)\|^2 ds$ is bounded (since t is fixed). Thus, for $\delta < 1$, one has

$$T_t U(M_2) - T_t U(M_1) \leq C\delta + \frac{1}{2\delta}\|M_2 - M_1\|^2 + \frac{C}{\sqrt{\delta}}\|M_2 - M_1\|$$

$$\leq C(\delta + \sqrt{\delta}) =: \varepsilon$$

whenever $\|M_2 - M_1\| \leq \delta$. One can now interchange the roles of M_1 and M_2 to conclude. $\qquad\square$

Since we took $c = 0$, we denote $U(\cdot; 0)$ and $u^n(\cdot; 0)$ by U and u^n, respectively. We know that $U(M_n) = u^n(x)$ whenever $M_n \equiv x_i$ is piecewise constant on the n-regular partition of $(0, 1)$. Therefore, it makes sense to write $T_t^n U(M_n)$ which is nothing but $T_t^n u^n(x)$.

Lemma 4.9. *If U is the one defined in Proposition 3.5, then for any $t > 0$ and any $M \in L^2(0, 1)$ there exists a sequence $M_n \in \mathcal{C}_n$ such that*

$$\|M_n - M\| \to 0 \text{ and } \limsup_{n \to \infty} T_t^n U(M_n) \leq T_t U(M). \qquad (4.9)$$

Proof. Let $\varepsilon_n \downarrow 0$. Since U is continuous and $\mathcal{C}$ is dense in $L^2(0, 1)$, for n sufficiently large we can find $M_n \in \mathcal{C}_n$ such that $\|M_n - M\| \leq \varepsilon_n/\kappa$. Then $|U(M_n) - U(M)| \leq \varepsilon_n$. Also, take $\gamma_n \in H^1(0, t; L^2(0, 1))$ such that $\gamma_n(t) = M$ and

$$\mathcal{A}(t; \gamma_n) - \varepsilon_n \leq T_t U(M) \leq U(\gamma_n(0))$$
$$+ \int_0^t L(\gamma_n(s), \dot{\gamma}_n(s)) ds =: \mathcal{A}(t; \gamma_n). \qquad (4.10)$$

Let $\sigma_n \in L^2(0, t; \mathcal{C}_n)$ such that $\|\dot{\gamma}_n - \sigma_n\|_{L^2((0,t) \times (0,1))} \leq \varepsilon_n/(\kappa\sqrt{t})$ (see, for example, [17] for the existence of such σ_n). Then place

$$S_n(s, x) = M_n(x) - \int_s^t \sigma_n(\tau, x) d\tau.$$

Obviously, $S_n \in H^1(0, t; \mathcal{C}_n)$. We have

$$S_n(s) - \gamma_n(s) = M_n - M + \int_s^t (\dot{\gamma}_n - \sigma_n) d\tau$$

which implies

$$\|S_n(s) - \gamma_n(s)\| \leq \frac{2\varepsilon_n}{\kappa} \text{ for } 0 \leq s \leq t. \qquad (4.11)$$

Since $S_n(t) = M_n$, we can write

$$
\begin{aligned}
T_t^n U(M_n) &\leq U(S_n(0)) + \int_0^t L(S_n(s), \sigma_n(s)) ds \\
&\leq U(S_n(0)) - U(\gamma_n(0)) + \mathcal{A}(t; \gamma_n) \\
&\quad + \int_0^t \left[L(S_n(s), \sigma_n(s)) - L(\gamma_n(s), \dot{\gamma}_n(s)) \right] ds \\
&\leq 3\varepsilon_n + T_t U(M) + \int_0^t \left[L(S_n(s), \sigma_n(s)) - L(\gamma_n(s), \dot{\gamma}_n(s)) \right] ds,
\end{aligned}
\tag{4.12}
$$

where we have taken into account the Lipschitz property of U (with Lipschitz constant at most κ) and (4.10). As for the last term in the right hand side, that is nothing but

$$
\frac{1}{2} \left[\|\sigma_n\|_{L^2(X_t)}^2 - \|\dot{\gamma}_n\|_{L^2(X_t)}^2 \right] - \frac{1}{2} \int_0^t \int_{(0,1)^2} \left[W(S_n(s, x) - S_n(s, y)) \right.
$$
$$
\left. - W(\gamma_n(s, x) - \gamma_n(s, y)) \right],
$$

where $X_t := (0, t) \times (0, 1)$. But (4.10) implies

$$
\frac{1}{2} \|\dot{\gamma}_n\|_{L^2(X_t)}^2 \leq T_t U(M) + \varepsilon_n + t \sup |W|,
$$

so $\|\dot{\gamma}_n\|_{L^2(X_t)}$ is bounded and, since $\|\dot{\gamma}_n - \sigma_n\|_{L^2(X_t)} \leq \varepsilon_n / (\kappa \sqrt{t})$, $\|\sigma_n\|_{L^2(X_t)}$ is also bounded. These considerations, along with (4.11) and the Lipschitz continuity of V, imply

$$
\lim_{n \to \infty} \int_0^t \left[L(S_n(s), \sigma_n(s)) - L(\gamma_n(s), \dot{\gamma}_n(s)) \right] ds = 0.
$$

Therefore, (4.12) yields the second statement in (4.9). $\qquad \square$

We have now all the tools to prove Proposition 4.6. Let us remind the reader that we do not lose generality by considering the case $c = 0$ only.

Proof of Proposition 4.6. Take $M \in L^2(0, 1)$ and the corresponding sequence M_n from Lemma 4.9. Note that, since $M_n \in C_n$ for all n, (4.7) enables us to write

$$
T_t^n U(M_n) = U(M_n) \to U(M) \text{ as } n \to \infty,
$$

where the convergence is due to the continuity of U. By using Lemma 4.9 again we pass to lim sup in the left hand side to deduce

$$
T_t U(M) \geq U(M).
$$

To prove the opposite inequality, note that

$$T_t U(M_n) \leq T_t^n U(M_n) \text{ for all } n \geq 1, \ t > 0$$

because the infimum in the definition of the left hand side is taken under fewer restrictions. Now Lemma 4.8 applies to yield the convergence of $T_t U(M_n)$ to $T_t U(M)$. But we have already seen that the right hand side converges to $U(M)$, so we conclude the proof. □

4.3 Forward semigroup

Define the (forward) semigroup $\tilde{T}_{L,t}$ on $C(\mathbb{T})$ by

$$\tilde{T}_{L,t} V(M) = \sup_{S(0)=M} \left\{ V(S(t)) - \int_0^t L(S(s), \dot{S}(s)) ds \right\}.$$

One can modify the proof of Proposition 4.6 to prove:

Proposition 4.10. *For any $c \in \mathbb{R}$ there exists a Lipschitz continuous, periodic, rearrangement invariant map $\tilde{U}(\cdot; c) : L^2(0, 1) \to \mathbb{R}$ such that*

$$\tilde{T}_{L_c,t}\tilde{U}(\cdot; c) = \tilde{U}(\cdot; c) + \frac{1}{2}c^2 t \text{ for all } t \geq 0. \tag{4.13}$$

Furthermore, $\tilde{U}(M_n; c) = \tilde{u}^n(x; c)$ whenever $M_n \in C_n$ and x is the corresponding vector in $\mathbb{R}^n$, where $\tilde{u}^n(\cdot; c)$ is a forward semigroup Weak KAM solution on $\mathbb{T}^n$, i.e.

$$\tilde{T}^n_{L_c^n,t}\tilde{u}^n(\cdot; c) = \tilde{u}^n(\cdot; c) + \frac{1}{2}c^2 t \text{ for all } t \geq 0. \tag{4.14}$$

Indeed, this is, in some sense, the dual of Proposition 4.6 as it uses the forward semigroup $\tilde{T}_t$ instead of the more usual Lax-Oleinik backward semigroup T_t to construct viscosity solutions for (1.15). The following result yields the second statement of Theorem 1.1.

Proposition 4.11. *Let $M \in \mathcal{M}$ be fixed. Then for every $c \in \mathbb{R}$ there exists a global extremal curve $S \in H^2(0, \infty; L^2(0, 1))$ such that $S(0) = M$, $S(t) \in \mathcal{M}$ for all $t \geq 0$, and*

$$\tilde{U}(S(t); c) - \tilde{U}(M; c) = \int_0^t L_c(S(s), \dot{S}(s)) ds + \frac{1}{2}c^2 t \text{ for all } t \geq 0. \tag{4.15}$$

Proof. Let $M_n \in C_n$ be nondecreasing and such that $M_n \to M$ in $L^2(0, 1)$. According to Proposition 4.10, the restriction $\tilde{u}^n$ of $\tilde{U}$ to C_n (or, equivalently, $\mathbb{R}^n$) is a Weak KAM solution for (4.14) on $\mathbb{T}^n$. in [12, Theorem 4.5.3] provides a global extremal $\{(\sigma_n(s), \dot{\sigma}_n(s))\}_{s \geq 0} \subset \mathbb{R}^n \times \mathbb{R}^n$ such that

$$\tilde{u}^n(\sigma_n(t); c) - \tilde{u}^n(x_n; c) = \int_0^t L_c^n(\sigma_n(s), \dot{\sigma}_n(s))ds$$
$$+ \frac{1}{2}c^2 t \text{ for all } t \geq 0, \tag{4.16}$$

where x_n is the n-dimensional vector corresponding to M_n. Note that the path $s \to \sigma_n(s)$ we consider here is, in fact, a lift of the one from [12] to the universal cover $\mathbb{R}^n$. So, if we denote by $S_n(s)$ the function in C_n corresponding to $\sigma_n(s) \in \mathbb{R}^n$, then we can write

$$\tilde{U}(S_n(t); c) - \tilde{U}(M_n; c) = \int_0^t L_c(S_n(s), \dot{S}_n(s))ds$$
$$+ \frac{1}{2}c^2 t \text{ for all } t \geq 0. \tag{4.17}$$

We first deduce

$$\tilde{U}(S_n(t); c) - \tilde{U}(M_n; c) \geq \frac{1}{2}\int_0^t \|\dot{S}_n(s) - c\|^2 ds \text{ for all } t \geq 0 \tag{4.18}$$

which means $S_n^c(s) := S_n(s) - cs$ has functional time derivative bounded in $L^2(0, \infty; L^2(0, 1))$ uniformly with respect to n. It follows that $t \to S_n^c(t)$ is uniformly Hölder. In particular, for each $t \geq 0$, $S_n^c(t)$ is bounded in $L^2(0, 1)$, uniformly with respect to n. Note that, since M_n is nondecreasing we may assume [15,17] that $S_n^c(t)$ is nondecreasing for all n and t. Therefore, for any $t \geq 0$, there exists a subsequence $n_k \to \infty$ such that $S_{n_k}^c(t) \to S^c(t)$ in $L_{\text{loc}}^2(0, 1)$ for some $S^c(t) \in \mathcal{M}$ (see, e.g., [17]). By a standard diagonalization argument, we can use the same subsequence for all $t \in [0, \infty) \cap \mathbb{Q}$. Again, by a standard argument, one notices that $t \to S^c(t)$ is Hölder continuous on $[0, \infty) \cap \mathbb{Q}$, so it can be extended to the whole $[0, \infty)$ in a unique way. Furthermore, after relabeling $S_{n_k}^c(t)$ by $S_n^c(t)$, we use the uniform Hölder continuity of S_n^c and S^c (as well as the density of $\mathbb{Q}$ in $\mathbb{R}$) to deduce that $S_n(t) \to S^c(t) + ct =: S(t)$ in $L_{\text{loc}}^2(0, 1)$ for all $t \geq 0$. But the uniform bound on $\dot{S}_n^c$ in $L^2(0, \infty; L^2(0, 1))$ and some of the considerations above, also imply that $S \in H^1(0, \infty; L^2(0, 1))$ and, up to an unrelabeled subsequence, $\dot{S}_n \rightharpoonup \dot{S}$ weakly in $L^2(0, \infty; L^2(0, 1))$. Now let us assume $\tilde{U}(\cdot; c)$ is continuous with respect to the $L_{\text{loc}}^2(0, 1)$-topology as well (stronger than the already known $L^2(0, 1)$-continuity). If we pass to liminf in (4.17) as $n \to \infty$, we obtain (due to L_c being lower semicontinuous)

$$\tilde{U}(S(t); c) - \tilde{U}(M; c) \geq \int_0^t L_c(S(s), \dot{S}(s))ds + \frac{1}{2}c^2 t \text{ for all } t \geq 0$$

which, in light of (4.13) and the definition of $\tilde{T}_{L_c,t}$, turns into the equality (4.15). To prove the L^2_{loc}-continuity of $\tilde{U}(\cdot; c)$ let us take $\Omega \subset\subset (0, 1)$, $f \in L^2(0, 1)$ and simply remark that

$$\left|\tilde{U}(f\chi_\Omega) - \tilde{U}(f)\right| = \left|\tilde{U}(\widehat{f\chi_\Omega}) - \tilde{U}(\hat{f})\right| = \left|\tilde{U}(\hat{f}\chi_\Omega) - \tilde{U}(\hat{f})\right|$$

$$\leq \mathrm{Lip}(\tilde{U}(\cdot; c)) \left(\int_{(0,1)\setminus\Omega} |\hat{f}(x)|^2 dx\right)^{1/2}$$

$$\leq \mathrm{Lip}(\tilde{U}(\cdot; c))\mathcal{L}^1((0, 1)\setminus\Omega),$$

where we remind the reader that $\hat{f} = f - [f]$ (here $[\cdot]$ stands for the integer part function). Thus, the Lipschitz continuity of $\tilde{U}(\cdot; c)$ with respect to the L^2-topology implies its uniform continuity with respect to the L^2_{loc}-topology. $\qquad\square$

5   A spatially periodic Vlasov-Poisson system

We claim that Proposition 4.6 holds even if W is not necessarily C^2. The need for this regularity assumption came from the use of Theorem 4.5.3 in [12] to provide us with a global extremal $\{(\sigma_n(s), \dot{\sigma}_n(s))\}_{s\geq 0} \subset \mathbb{R}^n \times \mathbb{R}^n$ satisfying (4.16), and it has no bearing on Proposition 4.6. Indeed, we proved Proposition 4.6 by approximation with finite-dimensional Weak KAM solutions $u^n(\cdot; c)$ which were extracted from [5] (therefore, independently of [12]). To further explain, we refer back to Remark 3.2 (where we indicate how our $u^n(\cdot; c)$ is constructed) and point to the conditions on the Hamiltonian (1), (2) and (3) in [5]. These conditions will still be satisfied by H^n defined in (3.2) if W is only Lipschitz continuous (instead of C^2) and $\mathbb{Z}$-periodic. If (1.13) is further assumed on W, then the entire construction from Section 3 carries through. Lemmas 4.8 and 4.9 will also hold, so Proposition 4.6 will remain true in this less regular case.

In order to be able to reproduce the proof of Proposition 4.11 in this case, it would suffice to know that the global extremals $\{(\sigma(s), \dot{\sigma}(s))\}_{s\geq 0} \subset \mathbb{R}^n \times \mathbb{R}^n$ employed in (4.16) still exist at the n-dimensional level. For that, we apply a standard compactness argument in $H^1(0, t; \mathbb{R}^n)$ for a maximizing sequence in

$$
u^n(x; c) = \sup_{\sigma(0)=x} \left\{ u^n(\sigma(t); c) - \int_0^t L_c^n(\sigma(s), \dot{\sigma}(s))ds \right\} - \frac{1}{2}c^2 t,
$$

say it satisfies

$$
u^n(x; c) - \frac{1}{m} \le u^n(\sigma_m(t); c) - \int_0^t L_c^n(\sigma_m(s), \dot{\sigma}_m(s))ds - \frac{1}{2}c^2 t. \quad (5.1)
$$

The maximizing sequence $\{\sigma_m\}_m$ is bounded in $H^1(0, t; \mathbb{R}^n)$ (because $u^n(\cdot; c)$ is bounded in L^∞) and, since $\sigma_m(0) = x$ for all m we infer that, at least up to a subsequence, σ_m converges to some σ uniformly on $[0, t]$ (in particular, $\sigma(0) = x$) while $\dot{\sigma}_m$ converges to $\dot{\sigma}$ weakly in $L^2(0, t; \mathbb{R}^n)$. We obtain the desired result by passing to liminf as $m \to \infty$ in (5.1). Thus, Theorem 1.1 and Corollary 1.3 remain true for W only Lipschitz continuous. Not the same can be said about Corollary 1.2, as W is not regular enough to define the flow Ψ.

A periodic version of the Vlasov-Poisson system is replacing (1.8) if the C^2 potential W is replaced by the less regular

$$
W(z) = \frac{1}{2}\left(|z|_{\mathbb{T}^1}^2 - |z|_{\mathbb{T}^1}\right). \quad (5.2)
$$

Indeed, one checks by direct computation that the convolution $P := W * \rho$ satisfies

$$
1 - \partial_{xx}^2 P = \sum_{k \in \mathbb{Z}} \rho(\cdot + k). \quad (5.3)
$$

Consequently, with the potential (5.2) the system (1.8) turns into

$$
\begin{cases}
\partial_t f_t + v\partial_x f_t = \partial_x P_t \, \partial_v f_t \\
1 - \partial_{xx}^2 P_t = \sum_{k \in \mathbb{Z}} \rho_t(\cdot + k) \\
\rho_t = \int_{\mathbb{R}} f_t dv.
\end{cases} \quad (5.4)
$$

Any Borel probability μ^* on $\mathbb{T}^1$ can be represented as the $\mathbb{Z}$-indexed sum of integer translations of a Borel probability μ supported in $[0, 1)$. Thus, in light of (5.3) we deduce that

$$
P_{\mu^*}(x) = \int_{\mathbb{R}} W(x - z)d\mu(z) = \int_{\mathbb{T}^1} W(x - z)d\mu^*(z)
$$

satisfies

$$
1 - \partial_{xx}^2 P_{\mu^*} = \mu^* \quad \text{for all } \mu^* \in \mathcal{P}(\mathbb{T}^1).
$$

Furthermore, note that if $[0, \infty) \ni t \to f_t \in \mathcal{P}_2(\mathbb{R} \times \mathbb{R})$ satisfies (5.4) in the sense of distributions, then so does $t \to f_t(\cdot + k, \cdot)$ for any $k \in \mathbb{Z}$. By the linearity in f_t of the first equation in (5.4), we deduce that

$$f_t^* := \sum_{k \in \mathbb{Z}} f_t(\cdot + k, \cdot) \in \mathcal{P}(\mathbb{T}^1 \times \mathbb{R}) \tag{5.5}$$

satisfies (in the sense of distributions)

$$\begin{cases} \partial_t f_t^* + v \partial_x f_t^* = \partial_x P_t \, \partial_v f_t^* \\ 1 - \partial_{xx}^2 P_t = \rho_t^* \qquad \qquad \text{in } \mathbb{T}^1 \times \mathbb{R}. \\ \rho_t^* = \displaystyle\int_{\mathbb{R}} f_t^* dv \end{cases} \tag{5.6}$$

Let $c \in \mathbb{R}$ and $\mu^* \in \mathcal{P}(\mathbb{T}^1)$. We now have all the ingredients for a proof of the following:

Theorem 5.1. *There exists a path* $[0, \infty) \ni t \to f_t^* \in \mathcal{P}_2(\mathbb{T}^1 \times \mathbb{R})$ *satisfying in the distributional sense the spatially periodic Vlasov-Poisson system* (5.6) *with* $\rho_0^* = \mu^*$ *and such that*

$$\sup_{t>0} \sqrt{t} \, \|\mathbf{id}/t - c\|_{\rho_t} < \infty, \qquad \lim_{t \to \infty} \int_{\mathbb{T}^1} \int_{\mathbb{R}} |v - c|^2 d f_t^*(x, v) = 0,$$

where $\rho_t \in \mathcal{P}_2(\mathbb{R})$ *is such that*

$$\rho_t^* = \sum_{k \in \mathbb{Z}} \rho_t(\cdot + k).$$

Proof. Let $\mu \in \mathcal{P}([0, 1)) \subset \mathcal{P}_2(\mathbb{R})$ such that

$$\mu^* = \sum_{k \in \mathbb{Z}} \mu(\cdot + k).$$

According to Corollary 1.3 and our observations above for less regular potentials, there is a path $t \to \rho_t \in AC_{\text{loc}}^2(0, \infty; \mathcal{P}_2(\mathbb{R}))$ and $u : (0, \infty) \times \mathbb{R} \to \mathbb{R}$ Borel satisfying $u_t \in L^2(\rho_t)$ for $\mathcal{L}^1$-almost every $t > 0$, and $\rho_0 = \mu$. Also, (1.16) is satisfied with the third equation replaced by the second equation in (5.4). Set

$$f_t(x, v) := \rho_t(x)\delta_{u_t(x)}(v) \in \mathcal{P}_2(\mathbb{R} \times \mathbb{R}).$$

This path $t \to f_t$ satisfies (5.4) in the sense of distributions, and we have already shown above that f_t^* given by (5.5) solves (5.6) in the distributional sense with $\rho_0^* = \mu^*$. The asymptotic statement on the energy of f_t^* follows from the second equation in (1.17). $\qquad \square$

References

[1] L. AMBROSIO and W. GANGBO, *Hamiltonian ODEs in the Wasserstein Space of Probability Measures*, Comm. Pure Appl. Math. **LXI** (2008), 18–53.

[2] L. AMBROSIO, N. GIGLI and G. SAVARÉ, "Gradient flows in metric spaces and the Wasserstein spaces of probability measures", Lectures in mathematics, E.T.H. Zurich, Birkhäuser, 2005.

[3] W. BRAWN and K. HEPP, *The Vlasov dynamics and its fluctuations in the 1/N limit of interacting classical particles*, Comm. Math. Phys. **56** (1977), 101–113.

[4] H. BREZIS, "Analyse fonctionnelle; théorie et applications", Masson, Paris, 1983.

[5] M. C. CONCORDEL. *Periodic homogenization of Hamilton-Jacobi equations: additive eigenvalues and variational formula*, Indiana Univ. Math. J. **45** (1996), 1095–1117.

[6] M. G. CRANDALL and P. L. LIONS, *Hamilton-Jacobi equations in infinite dimensions I. Uniqueness of viscosity solutions*, J. Funct. Anal. **62** (1985), 379–396.

[7] M. CRANDALL and P. L. LIONS, *Hamilton-Jacobi equations in infinite dimensions II. Existence of viscosity solutions*, J. Funct. Anal. **65** (1986), 368–405.

[8] M. CRANDALL and P. L. LIONS, *Hamilton-Jacobi equations in infinite dimensions III*, J. Funct. Anal. **68** (1986), 214–247.

[9] R. L. DOBRUSHIN, *Vlasov equations*, Funct. Anal. Appl. **13** (1979), 115–123.

[10] W. E, *A class of homogenization problems in the Calculus of Variations*, Comm. Pure Appl. Math. **XLIV** (1991), 733–759.

[11] L. C. EVANS, *A survey of partial differential equations methods in weak KAM theory*, Comm. Pure Appl. Math. **57** (2004).

[12] A. FATHI, "Weak KAM theory in Lagrangian dynamics", preliminary version, Lecture notes, 2003.

[13] W. H. FLEMING and H. M. SONER, "Controlled Markov processes and viscosity solutions", Springer, New York, 1993.

[14] W. GANGBO, H. K. KIM and T. PACINI, *Differential forms on Wasserstein space and infinite-dimensional Hamiltonian systems*, to appear in Memoirs of AMS.

[15] W. GANGBO, T. NGUYEN and A. TUDORASCU, *Euler-Poisson systems as action minimizing paths in the Wasserstein space*, Arch. Rat. Mech. Anal. **192** (2008), 419–452.

[16] W. GANGBO and A. TUDORASCU, *Lagrangian Dynamics on an infinite-dimensional torus; a Weak KAM theorem*, preprint.

[17] W. GANGBO and A. TUDORASCU, *Homogenization for Hamilton-Jacobi equations in probability spaces*, in progress.

[18] P. L. LIONS, G. PAPANICOLAOU and S. R. S. VARADHAN, *Homogenization of Hamilton-Jacobi equations*, unpublished, cca 1988.

[19] V. P. MASLOV, "Self-consistent field equations", Contemporary Problems in Mathematics, Vol. 11, VINITI, Moscow, 1978, 153–234.

[20] J. MATHER, *Minimal measures*, Comment. Math. Helv. **64** (1989), 375–394.

[21] J. MATHER, *Action minimizing invariant measures for positive definite Lagrangian systems*, Math. Z. **207** (1991), 169–207.

[22] J. MATHER, *Variational construction of connecting orbits*, Ann. Inst. Fourier (Grenoble) **43** (1993), 1349–1386.

[23] H. R. MORTON, *Symmetric products of the circle*, Proc. Cambridge Philos. Soc. **63** (1967), 349–352.

[24] T. NGUYEN and A. TUDORASCU, *Pressureless Euler/Euler-Poisson systems via adhesion dynamics and scalar conservation laws*, SIAM J. Math. Anal. (to appear).

[25] E. SPANIER, *Infinite symmetric products, function spaces, and duality*, Annals of Mathematics **1** (1959), 142–198.

[26] P. VANHAECKE, *Integrable systems and symmetric products*, Mathematische Zeitschritft **227** (1998), 93–127.

[27] C. VILLANI, *Topics in optimal transportation*, Graduate Studies in Mathematics, American Mathematical Society **58** (2003).

Symmetrization, optimal transport and quantitative isoperimetric inequalities

Francesco Maggi

Abstract. In these lecture notes we introduce some recent developments on stability theorems for isoperimetric and Sobolev type inequalities. We discuss two different approaches to the stability problem. The first approach moves from some classical inequalities in symmetrization theory, while the second approach is based on mass transportation theory. We illustrate these two methods on the fundamental examples provided by the Euclidean isoperimetric inequality and by the Wulff inequality. A final section is devoted to an overview on further related problems.

1 The Euclidean isoperimetric inequality and symmetrization

Let E be an open bounded set with C^1-boundary in $\mathbb{R}^n$, $n \geq 2$. We denote by $|E|$ the Lebesgue measure of E and define the perimeter $P(E)$ of E as the $(n-1)$-dimensional Hausdorff measure $\mathcal{H}^{n-1}(\partial E)$ of its topological boundary ∂E. We have then the *Euclidean isoperimetric inequality*,

$$P(E) \geq n|B|^{1/n}|E|^{(n-1)/n}, \tag{1.1}$$

where B denotes the Euclidean unit ball centered at the origin. Since $P(B) = n|B|$, if $E = B(x,r) = x + rB$ for some $x \in \mathbb{R}^n$ and $r > 0$, then equality holds in (1.1). The converse is also true: if equality holds in (1.1), then there exist $x \in \mathbb{R}^n$ and $r > 0$ such that $E = B(x,r)$. Therefore the Euclidean isoperimetric inequality – together with a characterization of its equality cases – provides a synthetic formulation of the isoperimetric theorem: among sets of given Lebesgue measure, balls (and only balls) minimize the perimeter. This section is devoted to a detailed sketch of the classical proof of the Euclidean isoperimetric inequality by symmetrization theory. We follow closely the argument in [18] (see also [27] and [40, Section 3.1-3.2]). The first step of the proof consists in showing the existence of an optimal set for the variational problem,

$$\inf\{P(E) : |E| = |B|\}. \tag{1.2}$$

It is absolutely non trivial to prove this statement by bare hands. This is the kind of assertion that becomes standard once the theory of sets of finite perimeter is at disposal, and it is actually in this framework that most of the ideas presented in these notes can be turned into rigorous proofs. However, in order to make our exposition more accessible, we avoid any reference to geometric measure theory. In particular, we shall take for granted the existence of a minimizer in the variational problem (1.2) and, even if this minimizer is in principle just a set of finite perimeter, we are going to treat it as if it where an open set with C^1-boundary or with polyhedral boundary, depending on our convenience.

To show that every minimizer in (1.2) is a ball we introduce the notion of Steiner symmetrization. Let us decompose $x \in \mathbb{R}^n$ as $x = (x', x_n) \in \mathbb{R}^{n-1} \times \mathbb{R}$. Correspondingly, given $E \subset \mathbb{R}^n$ we define a function $m : \mathbb{R}^{n-1} \to \mathbb{R}$ on setting

$$m(x') = \mathcal{H}^1(\{t \in \mathbb{R} : (x', t) \in E\}), \quad x' \in \mathbb{R}^{n-1},$$

and then we use m to define E^s, the *Steiner symmetrization* of E, as

$$E^s = \left\{ x \in \mathbb{R}^n : |x_n| < \frac{m(x')}{2} \right\}.$$

Thus, E^s is obtained by rearranging the one-dimensional vertical slices

$$E(x') = \{t \in \mathbb{R} : (x', t) \in E\}, \quad x' \in \mathbb{R}^{n-1},$$

into vertical segments centered at $(x', 0)$. We notice that E^s is symmetric by reflection with respect to $\mathbb{R}^{n-1}$ and that, by Fubini's theorem, $|E| = |E^s|$. Moreover, the perimeter is lowered under Steiner symmetrization, *i.e.* we have the so called *Steiner inequality*,

$$P(E) \geq P(E^s). \tag{1.3}$$

The proof of the first two facts is easy. We now discuss a proof of (1.3).

Sketch of the proof of (1.3). We directly consider the case that E is a bounded set with polyhedral boundary. Since the outer unit normal ν_E of E takes finitely many values, up to an arbitrarily small rotation of E we may assume that $\nu_E \cdot e_n \neq 0$ everywhere. As a consequence we can completely parameterize ∂E as the union of an even number of graphs of piecewise affine functions.

More precisely – see Figure 1.1 – if E' is the projection of E over $\mathbb{R}^{n-1}$ (note that E' is a polyhedral set in $\mathbb{R}^{n-1}$), then there exist $N \in \mathbb{N}$, a partition $\{D_h\}_{h=1}^N$ of E' into polyhedral sets in $\mathbb{R}^{n-1}$, and affine functions $f_{h,k}, g_{h,k} : D_h \to \mathbb{R}, 1 \le k \le M(h)$, such that

$$E \cap (D_h \times \mathbb{R}) = \bigcup_{k=1}^{M(h)} \{(x', x_n) : x' \in D_h, g_{h,k}(x') < x_n < f_{h,k}(x')\}.$$

By construction we have that

$$m(x') = \sum_{k=1}^{M(h)} (f_{h,k}(x') - g_{h,k}(x')), \quad x' \in D_h. \tag{1.4}$$

Hence,

$$\mathcal{H}^{n-1}((D_h \times \mathbb{R}) \cap \partial E^s) = 2 \int_{D_h} \sqrt{1 + |\nabla m/2|^2} = \int_{D_h} \sqrt{4 + |\nabla m|^2},$$

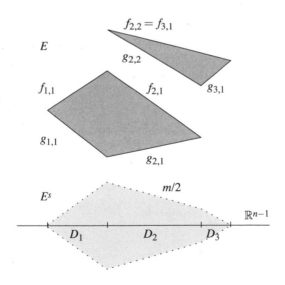

Figure 1.1. The parameterization of E used in the proof of Steiner inequality. In this case $M(1) = M(3) = 1$ and $M(2) = 2$.

for every $1 \leq h \leq N$ and, moreover, by Jensen's inequality and by (1.4),

$$
\mathcal{H}^{n-1}((D_h \times \mathbb{R}) \cap \partial E) = \sum_{k=1}^{M(h)} \int_{D_h} \sqrt{1 + |\nabla f_{h,k}|^2} + \sqrt{1 + |\nabla g_{h,k}|^2}
$$

$$
\geq 2 \sum_{k=1}^{M(h)} \int_{D_h} \sqrt{1 + \left| \frac{\nabla f_{h,k} - \nabla g_{h,k}}{2} \right|^2}
$$

$$
\geq 2M(h) \int_{D_h} \sqrt{1 + \left| \frac{1}{M(h)} \sum_{k=1}^{M(h)} \frac{\nabla f_{h,k} - \nabla g_{h,k}}{2} \right|^2}
$$

$$
= \int_{D_h} \sqrt{4M(h)^2 + |\nabla m|^2}.
$$

Since $M(h) \geq 1$ we have thus proved that

$$
\mathcal{H}^{n-1}((D_h \times \mathbb{R}) \cap \partial E) \geq \mathcal{H}^{n-1}((D_h \times \mathbb{R}) \cap \partial E^s).
$$

Adding up over $1 \leq h \leq N$ we deduce (1.3). □

Remark 1.1. If equality holds in (1.3), then we have $M(h) = 1$ for every h, *i.e.*, the vertical sections of E over E' are segments (to be more precise: for $\mathcal{H}^{n-1}$-a.e. $x' \in E'$ we have that $\{t \in \mathbb{R} : (x', t) \in E\}$ is $\mathcal{H}^1$-equivalent to a segment).

Remark 1.2. If E is convex and equality holds in (1.3), then E is a vertical translation of E^s, *i.e.* $E = s e_n + E^s$ for some $s \in \mathbb{R}$. Indeed if E is convex then the projection E' of E over $\mathbb{R}^{n-1}$ is a convex set and there exist two concave functions $\psi_1, \psi_2 : E' \to \mathbb{R}$ such that $\psi_1 + \psi_2 > 0$ on E' and

$$
E = \{x \in \mathbb{R}^n : -\psi_1(x') < x_n < \psi_2(x'), x' \in E'\}.
$$

Then we have $E^s = \{x \in \mathbb{R}^n : 2|x_n| < \psi_1(x') + \psi_2(x'), x' \in E'\}$ and in particular

$$
P(E) = \int_{E'} \sqrt{1 + |\nabla \psi_1|^2} d\mathcal{H}^{n-1} + \int_{E'} \sqrt{1 + |\nabla \psi_2|^2} d\mathcal{H}^{n-1}
$$
$$
+ \int_{\partial E'} |\psi_1 + \psi_2| d\mathcal{H}^{n-2},
$$

$$
P(E^s) = 2 \int_{E'} \sqrt{1 + \left| \frac{\nabla \psi_1 + \nabla \psi_2}{2} \right|^2} d\mathcal{H}^{n-1} + \int_{\partial E'} |\psi_1 + \psi_2| d\mathcal{H}^{n-2}.
$$

We see again that $P(E) \geq P(E^s)$ thanks to Jensen's inequality. Since the function $\sqrt{1+z^2}$ is strictly convex on $z \in \mathbb{R}^n$ we actually have that $P(E) = P(E^s)$ implies $\nabla \psi_1 = \nabla \psi_2$ on E' and therefore, since E' is connected, $\psi_1 = s + \psi_2$ for some constant $s \in \mathbb{R}$.

Sketch of the proof that a minimizer for (1.2) *is a ball.* Let now E be a minimizer in (1.2). We want first to prove that E is convex. To this end, let $x, y \in E$. Up to a rotation, that does not affect the minimality of E, we can assume that the segment $[x, y]$ is parallel to e_n. Let then E^s be the Steiner symmetrization of E. Since $|E^s| = |E|$, by minimality we have $P(E^s) \geq P(E)$. The reverse inequality is the Steiner inequality. Therefore $P(E^s) = P(E)$, and the vertical sections of E are segments by Remark 1.1. In particular, the segment $[x, y]$ is contained in E. The convexity of E is thus proved.

Since E is convex and $P(E) = P(E^s)$, by Remark 1.2 we have that, up to a vertical translation, $E = E^s$. We can repeat this argument on the other $(n-1)$-coordinate directions (as the various translations performed are mutually orthogonal). In this way we have proved that E is symmetric with respect the coordinate hyperplanes. In particular, E is symmetric by reflection through the origin, *i.e.* $x \in E$ if and only if $-x \in E$. Let now $e \in S^{n-1}$, then again there exists $\alpha \in \mathbb{R}$ such that E is symmetric by reflection with respect to the affine plane $\{x : x \cdot e = \alpha\}$. Since E is symmetric by reflection through the origin, however, it must be $\alpha = 0$. Thus E is symmetric by reflection with respect to every hyperplane *passing through the origin*. Let now $x \in E$ and let $y \in \mathbb{R}^n$ such that $|x| = |y|$. On reflecting E with respect to the hyperplane orthogonal to $y - x$ we see that $y \in E$. Since E is convex and contains the origin, $E = B$. $\square$

Remark 1.3 (Functional form of the Euclidean isoperimetric inequality).

The Euclidean isoperimetric inequality is in fact equivalent to the *Sobolev inequality for functions of bounded variation*,

$$|Du|(\mathbb{R}^n) \geq n|B|^{1/n} \|u\|_{L^{n/(n-1)}(\mathbb{R}^n)}, \quad u \in BV(\mathbb{R}^n), \qquad (1.5)$$

where equality holds if and only if u is a multiple of the characteristic function of a ball $B(x, r)$. Let us briefly justify these assertions. To begin with, we recall the definition of *total variation* over $\mathbb{R}^n$ for a function $u \in L^1_{\text{loc}}(\mathbb{R}^n)$, namely we set

$$|Du|(\mathbb{R}^n) = \sup\left\{\int_{\mathbb{R}^n} u(x)\,\mathrm{div}\,T(x)\,dx : T \in C^1_c(\mathbb{R}^n; \mathbb{R}^n), |T| \leq 1\right\}.$$

Correspondingly, we introduce the space of functions of bounded variation over $\mathbb{R}^n$

$$BV(\mathbb{R}^n) = \{u \in L^1(\mathbb{R}^n) : |Du|(\mathbb{R}^n) < \infty\}.$$

For the reader who is unfamiliar with functions of bounded variation let us just notice that $W^{1,1}(\mathbb{R}^n) \subset BV(\mathbb{R}^n)$ with

$$|Du|(\mathbb{R}^n) = \int_{\mathbb{R}^n} |\nabla u(x)|dx, \quad \forall u \in W^{1,1}(\mathbb{R}^n). \tag{1.6}$$

At the same time, by the divergence theorem, if $u = 1_E$ is the characteristic function of an open bounded set E with C^1-boundary then $u \in BV(\mathbb{R}^n)$ and

$$|Du|(\mathbb{R}^n) = \mathcal{H}^{n-1}(\partial E) = P(E).$$

In particular, if we plug $u = 1_E$ into (1.5) then we find the Euclidean isoperimetric inequality (1.1). To show the converse implication let us first recall a basic property of BV-functions, namely the coarea formula

$$|Du|(\mathbb{R}^n) = \int_{\mathbb{R}} P(\{u > t\})\,dt. \tag{1.7}$$

We briefly justify (1.7) by proving it in the case that u is a piecewise affine function belonging to $W^{1,1}(\mathbb{R}^n)$ (the general case follows then by a standard approximation argument). For every such u there exist a partition of $\mathbb{R}^n$ into Borel sets $\{E_i\}_{i=1}^N$, $c_i \in \mathbb{R}$ and $v_i \in \mathbb{R}^n$ such that $u(x) = c_i + v_i \cdot x$ for $x \in E_i$. Therefore by (1.6),

$$|Du|(\mathbb{R}^n) = \sum_{i=1}^N |v_i|\,|E_i|.$$

By Fubini's theorem and by a change of variable we find that

$$|v_i|\,|E_i| = |v_i| \int_{\mathbb{R}} \mathcal{H}^{n-1}\left(E_i \cap \left\{x \cdot \frac{v_i}{|v_i|} = s\right\}\right)ds$$

$$= \int_{\mathbb{R}} \mathcal{H}^{n-1}(E_i \cap u^{-1}\{t\})\,dt.$$

Hence, adding up over i, we prove (1.7)

$$|Du|(\mathbb{R}^n) = \int_{\mathbb{R}} \mathcal{H}^{n-1}(u^{-1}\{t\})dt = \int_{\mathbb{R}} P(\{u > t\})\,dt.$$

We now pass to deduce (1.5) from (1.1). Let us assume for the sake of clarity that $u \in C_c^\infty(\mathbb{R}^n)$ with $u \geq 0$. We may apply the Euclidean isoperimetric inequality to each level set $\{u > t\}$ into the coarea formula, so to find that

$$|Du|(\mathbb{R}^n) \geq n|B|^{1/n} \int_0^\infty \mu(t)^{(n-1)/n} dt, \quad \mu(t) = |\{u > t\}|. \quad (1.8)$$

If we take the derivative of the increasing function

$$\varphi(s) = \left(\int_0^s \mu(t)^{(n-1)/n} dt \right)^{n/(n-1)},$$

and notice that μ is decreasing on $[0, \infty)$ then we see that

$$\varphi'(s) = \frac{n}{n-1} \left(\int_0^s \mu(t)^{(n-1)/n} dt \right)^{1/(n-1)} \mu(s)^{(n-1)/n}$$

$$\geq \frac{n}{n-1} \left(s \, \mu(s)^{(n-1)/n} \right)^{1/(n-1)} \mu(s)^{(n-1)/n} = \frac{n}{n-1} s^{1/(n-1)} \mu(s).$$

Therefore

$$\varphi(\infty) \geq \int_0^\infty \varphi'(s) ds \geq \frac{n}{n-1} \int_0^\infty s^{1/(n-1)} \mu(s) ds = \int_{\mathbb{R}^n} |u(x)|^{n/(n-1)} dx,$$

where the last equality follows again from a change of variables and Fubini's theorem. Thus

$$\int_0^\infty \mu(t)^{(n-1)/n} dt \geq \|u\|_{L^{n/(n-1)}(\mathbb{R}^n)}, \quad (1.9)$$

and (1.5) follows by combining (1.8) and (1.9). The above argument can be repeated on a non-negative function $u \in BV(\mathbb{R}^n)$. Equality in (1.5) then implies equality in (1.8) - so that $P(\{u > t\}) = n|B|^{1/n} \mu(t)^{(n-1)/n}$ for a.e. $t > 0$, i.e. $\{u > t\}$ is equivalent to a ball for a.e. $t \in \mathbb{R}$ - and it implies equality in (1.9), so that $\mu(t)$ is constant for $t \in (0, \|u\|_{L^\infty(\mathbb{R}^n)})$. Thus $u = a \, 1_{B(x,r)}$ for some $a > 0$, $x \in \mathbb{R}^n$ and $r > 0$. Moreover, by an analogous argument to that presented in Remark 2.2 we can see that if $u \in BV(\mathbb{R}^n)$ realizes equality in (1.5) then either $u \geq 0$ or $u \leq 0$.

2 The Wulff inequality and mass transport

The Euclidean isoperimetric inequality can be seen as a particular case of a more general variational principle, known as the Wulff inequality. The Wulff inequality originates from the physical problem of determining the equilibrium shape of a crystal in the absence of external forces. It is assumed that the equilibrium configuration is reached by minimizing (under a volume constraint) an anisotropic surface energy

$$\mathcal{F}(E) = \int_{\partial E} f(\nu_E) d\mathcal{H}^{n-1},$$

defined by a surface tension $f : \mathbb{R}^n \to [0, \infty)$. The surface tension f, that is determined by the physical properties of the crystal itself, is generally assumed to be convex and positively homogeneous of degree one, i.e. $f(t\nu) = t f(\nu)$ for $t > 0$ and $\nu \in S^{n-1}$. Typically, it is realized as the maximum of finitely many linear functions on $\mathbb{R}^n$. The Wulff shape associated to f is the open, bounded convex set

$$K = \bigcap_{\nu \in S^{n-1}} \{y \in \mathbb{R}^n : y \cdot \nu < f(\nu)\}.$$

We have then the *Wulff inequality*

$$\mathcal{F}(E) \geq n|K|^{1/n}|E|^{(n-1)/n}. \tag{2.1}$$

Since $\mathcal{F}(K) = n|K|$, we have equality in (2.1) when $E = x + r K$, for some $x \in \mathbb{R}^n$ and $r > 0$. Viceversa, it can be proved that if equality holds in (2.1) then $E = x + rK$ for some $x \in \mathbb{R}^n$ and $r > 0$ ([24,51]). In the particular case that $f(\nu) = |\nu|$, the Wulff shape reduces to the Euclidean unit ball, i.e. $K = B$, and $\mathcal{F}(E) = P(E)$. Thus we can see the Euclidean isoperimetric inequality as a particular case of the Wulff inequality.

Remark 2.1. In the discussion above we have constructed the Wulff shape K on starting from a given positively homogenous of degree one, non negative convex function f. It is sometimes useful to invert this operation. Given a bounded convex set K – possibly *not* containing the origin – we define a homogenous of degree one convex function $f_K : \mathbb{R}^n \to \mathbb{R}$ by setting

$$f_K(t \nu) = t \sup\{x \cdot \nu : x \in K\}, \quad t \geq 0, \nu \in S^{n-1}, \tag{2.2}$$

and we consider the surface energy functional $\mathcal{F}_K$ defined by

$$\mathcal{F}_K(E) = \int_{\partial E} f_K(\nu_E) d\mathcal{H}^{n-1}.$$

If $0 \in K$ then $f_K(\nu) > 0$ for every $\nu \in S^{n-1}$, and we are precisely in the framework described above: in particular, K is the Wulff shape corresponding to $\mathcal{F}_K$. If K does not contain the origin then f_K is not positive on S^{n-1}. However we can easily reduce to the previous case by noticing that for some $x_0 \in \mathbb{R}^n$ the bounded open convex set $x_0 + K$ contains the origin and that, correspondingly

$$f_{x_0+K}(\nu) = x_0 \cdot \nu + f_K(\nu),$$

for every $\nu \in S^{n-1}$. Since $\int_{\partial E} \nu_E \, d\mathcal{H}^{n-1} = 0$,

$$\mathcal{F}_{x_0+K}(E) = \int_{\partial E} (x_0 \cdot \nu_E) + f_K(\nu_E) \, d\mathcal{H}^{n-1} = \mathcal{F}_K(E),$$

and the surface energies of $\mathcal{F}_K$ and $\mathcal{F}_{x_0+K}$ are the same functional.

Starting from the work of Dinghas [16], the Wulff inequality is classically seen as a consequence of the *Brunn-Minkowski inequality*,

$$|E + F|^{1/n} \geq |E|^{1/n} + |F|^{1/n}, \tag{2.3}$$

where $E + F = \{x + y : x \in E, y \in F\}$ is the *Minkowski sum* of E and F. Indeed the anisotropic surface energy can be realized (on sets with sufficiently regular boundary) by a limiting procedure as

$$\mathcal{F}(E) = \lim_{\varepsilon \to 0^+} \frac{|E + \varepsilon K| - |E|}{\varepsilon}. \tag{2.4}$$

Then by (2.3) we have that

$$\mathcal{F}(E) \geq \lim_{\varepsilon \to 0^+} \frac{\psi(\varepsilon) - \psi(0)}{\varepsilon},$$

provided we set

$$\psi(\varepsilon) = (|E|^{1/n} + \varepsilon |K|^{1/n})^n - |E|, \quad \varepsilon > 0.$$

Since $\psi'(0) = n|K|^{1/n}|E|^{(n-1)/n}$, the Wulff inequality (2.1) follows immediately, although the passage to the limit prevents the possibility to characterize the equality cases by this argument. A striking proof of the Wulff inequality was obtained by Gromov [44] by means of the transport map between E and K known as the Knothe map [39]. We now sketch his argument.

Sketch of Gromov's proof of the Wulff inequality. Without loss of generality we assume that $|E| = |K|$ and that E and K have the same barycenter.

Step 1. We fix an orthonormal basis $\{v_k\}_{k=1}^n$ of $\mathbb{R}^n$ and denote by $x_k = x \cdot v_k, x \in \mathbb{R}^n$, the k-th coordinate function. We construct a map $M : E \to$

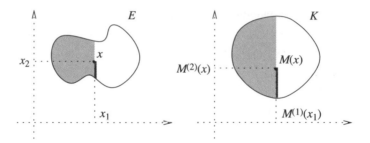

Figure 2.1. The construction of the Knothe map.

K such that ∇M is lower triangular (*i.e.*, the k-th component $M^{(k)} = v_k \cdot M$ of M depends on the first k-coordinates only), with

$$\lambda_k = \frac{\partial M^{(k)}}{\partial x_k} \geq 0, \quad 1 \leq k \leq n,$$

i.e. $M^{(k)}$ is non-decreasing in the k-th variable, and such that

$$\det \nabla M = \prod_{k=1}^{n} \lambda_k = 1 \quad \text{on } E.$$

For the sake of clarity we limit ourselves to describe this construction in the case $n = 2$, see Figure 2.1 (the generalization to higher dimension is however straightforward). Let $x \in E$. With a slight abuse of notation, we choose $M^{(1)}(x) = M^{(1)}(x_1)$ so that the fraction of the area of E contained in the half-space $\{z_1 < x_1\}$ is the same as the fraction of the area of K that is contained in the half-space $\{z_1 < M^{(1)}(x_1)\}$: *i.e.*, we set

$$\frac{|\{z \in E : z_1 < x_1\}|}{|E|} = \frac{|\{z \in K : z_1 < M^{(1)}(x_1)\}|}{|K|}. \tag{2.5}$$

In this way, every x belonging to the vertical section $E \cap \{z_1 = t\}$ of E is mapped by M into the vertical section $K \cap \{z_1 = M^{(1)}(t)\}$ of K. We then look at the fraction of the length of $E \cap \{z_1 = x_1\}$ that is contained in the half-space $\{z_2 < x_2\}$ and then choose $M^{(2)}(x_1, x_2)$ so that the same fraction of the length of $K \cap \{x_1 = M^{(1)}(x_1)\}$ is contained in the half-space $\{z_2 < M^{(2)}(x_1, x_2)\}$, *i.e.* we set

$$\begin{aligned} &\frac{\mathcal{H}^1(\{z \in E : z_1 = x_1, z_2 < x_2\})}{\mathcal{H}^1(\{z \in E : z_1 = x_1\})} \\ &= \frac{\mathcal{H}^1(\{z \in K : z_1 = M^{(1)}(x), z_2 < M^{(2)}(x)\})}{\mathcal{H}^1(\{z \in K : z_1 = M^{(1)}(x)\})}. \end{aligned} \tag{2.6}$$

We now see that M has the required properties. Clearly M maps E into K, $M^{(1)}$ does not depend on x_2 and $M^{(2)}$ is non-decreasing in x_2. By Fubini's theorem and by (2.5) we have that

$$\frac{1}{|E|} \int_{-\infty}^{x_1} dz_1 \int_{\mathbb{R}} 1_E(z) dz_2 = \frac{1}{|K|} \int_{-\infty}^{M^{(1)}(x_1)} dz_1 \int_{\mathbb{R}} 1_K(z) dz_2,$$

therefore by the chain rule

$$\frac{\partial M^{(1)}}{\partial x_1}(x_1) = \frac{|K|}{|E|} \frac{\mathcal{H}^1(\{z \in E : z_1 = x_1\})}{\mathcal{H}^1(\{z \in K : z_1 = M^{(1)}(x)\})} > 0. \qquad (2.7)$$

In the same way by (2.6) we have that

$$\frac{\int_{-\infty}^{x_2} 1_E(x_1, z_2) dz_2}{\mathcal{H}^1(\{z \in E : z_1 = x_1\})} = \frac{\int_{-\infty}^{M^{(2)}(x)} 1_K(M^{(1)}(x_1), z_2) dz_2}{\mathcal{H}^1(\{z \in K : z_1 = M^{(1)}(x_1)\})},$$

so that

$$\frac{\partial M^{(2)}}{\partial x_2}(x) = \frac{\mathcal{H}^1(\{z \in K : z_1 = M^{(1)}(x)\})}{\mathcal{H}^1(\{z \in E : z_1 = x_1\})} > 0. \qquad (2.8)$$

In particular, as $|E| = |K|$, we have that $\det \nabla M = 1$ on E.

Step 2. We now come to Gromov's argument. Let us first remark that we may associate to f a dual convex function $f_* : \mathbb{R}^n \to [0, \infty)$ so that

$$f_*(x) = \inf \left\{ \lambda > 0 : \frac{x}{\lambda} \in K \right\}.$$

Then f_* is convex, positively homogenous of degree one and such that

$$K = \{x \in \mathbb{R}^n : f_*(x) < 1\}, \qquad (2.9)$$
$$x \cdot y \le f(x) f_*(y), \quad \forall x, y \in \mathbb{R}^n. \qquad (2.10)$$

Let us now consider the Knothe map between E and K constructed in step one. Since $\lambda_k > 0$ whenever $1 \le k \le n$, then by the arithmetic-geometric mean inequality we have

$$1 = (\det \nabla M)^{1/n} = \prod_{k=1}^{n} \lambda_k^{1/n} \le \frac{1}{n} \sum_{k=1}^{n} \lambda_k = \frac{\text{div } M}{n}. \qquad (2.11)$$

Integrating this inequality over E and applying the divergence theorem we thus find that

$$n|K|^{1/n}|E|^{(n-1)/n} = n|E| = n \int_E (\det \nabla M)^{1/n}$$

$$\leq \int_E \operatorname{div} M = \int_{\partial E} M \cdot \nu_E \, d\mathcal{H}^{n-1}$$

$$\leq \int_{\partial E} f(\nu_E) f_*(M) d\mathcal{H}^{n-1}$$

$$\leq \mathcal{F}(E),$$

since $M(E) = K$ and $f_*(x) < 1$ if $x \in K$.

Step 3. We eventually characterize the equality cases in the Wulff inequality. If $\mathcal{F}(E) = n|K|^{1/n}|E|^{(n-1)/n}$ then equality holds in (2.11), so that there exists $\lambda : E \to \mathbb{R}$ such that $\lambda_k(x) = \lambda(x)$ for every $x \in E$. Since $1 = \det \nabla M = \lambda^n$ we actually find that $\lambda_k(x) = 1$ for every $x \in E$ and $1 \leq k \leq n$. This information is quite weak for $k \geq 2$, however when $k = 1$ it entails[1] the existence of $c \in \mathbb{R}$ such that

$$M^{(1)}(x_1) = x_1 + c, \quad x \in E.$$

From the definition (2.5) of $M^{(1)}$ (that is the same in every dimension) we see that

$$|\{z \in E : z_1 < x_1\}| = |\{z \in K : z_1 < x_1 + c\}|, \quad x \in E.$$

Since E and K have the same barycenter we deduce that $c = 0$. In particular the supporting half-space of K in the direction ν_1, i.e. $\{z : z_1 < f(\nu_1)\}$, contains E,

$$E \subset \{z : z_1 < f(\nu_1)\}.$$

We now repeat this argument in every direction. Given $\nu \in S^{n-1}$ we consider the supporting half-space to K in the ν-direction, $\{z : z \cdot \nu < f(\nu)\}$. Next we set $\nu_1 = \nu$ and consider $\{\nu_k\}_{k=2}^n$ so that $\{\nu_k\}_{k=1}^n$ is an orthonormal basis of $\mathbb{R}^n$. We correspondingly construct the Knothe map between E and K and argue as above to find that

$$E \subset \{z : z \cdot \nu < f(\nu)\}, \quad \forall \nu \in S^{n-1},$$

i.e. $E \subset K$. Since $|E| = |K|$ we have $E = K$.   □

---

[1] This is true provided E is connected, but optimal sets in the Wulff inequality are always connected, see Remark 2.2.

The use of infinitely many Knothe maps in the above argument should be compared with the use of Steiner symmetrization with respect to infinitely many directions in the proof of the Euclidean isoperimetric inequality. When studying stability issues we usually start from an argument characterizing the optimal sets and then try to turn the "exact" geometric information it provides into quantitative estimates. ¿From this point of view, the more direct the characterization argument, the stronger the stability estimate it leads to. Therefore, the use of infinitely many symmetrization steps/parametrization maps, if not really necessary, should lead to non-optimal estimates.

In the case of Gromov's proof of the Wulff inequality we can avoid the use of infinitely many Knothe maps by using an *optimal* mass transport map, the Brenier map. Under the assumption that $|E| = |K|$ we minimize the *transportation cost*

$$\int_E |T(x) - x|^2 dx, \tag{2.12}$$

among *transport maps* $T : \mathbb{R}^n \to \mathbb{R}^n$ between E and K, *i.e.* among maps such that

$$|E \cap T^{-1}(F)| = |F|, \quad \forall F \subset K. \tag{2.13}$$

By the change of variables,

$$|F| = \int_{T^{-1}(F)} |\det \nabla T| \, dx, \quad \forall F \subset K,$$

we see that a necessary and sufficient condition for T to be a transport map between E and K is that $T(E) \subset K$ and $|\det \nabla T| = 1$ a.e. on E. Therefore Knothe maps are transport maps between E and K. As shown by Brenier [8] (see also [43]) a transport map that minimizes the transportation cost (2.12) exists, it is unique (as a measurable map) and there exists a convex function $\varphi : \mathbb{R}^n \to \mathbb{R}$ such that $T = \nabla \varphi$ a.e. on $\mathbb{R}^n$. Note that, in dimension $n = 1$, this condition means of course that T is monotone increasing, and therefore the Knothe map (we have just one Knothe map when $n = 1$) and the Brenier map coincide. When $n \geq 2$, in most of the cases, the two constructions lead to very different transport maps.

Let now T be the Brenier map between E and K. Since $\nabla T = \nabla^2 \varphi$ and φ is convex, ∇T is symmetric and non-negative definite, *i.e.*

$$\nabla T(x) = \sum_{k=1}^n \lambda_k(x) e_k(x) \otimes e_k(x), \tag{2.14}$$

where $\{e_k(x)\}_{k=1}^n$ is an orthonormal basis of $\mathbb{R}^n$ and where $\lambda_k : \mathbb{R}^n \to [0, \infty)$, $1 \leq k \leq n$. Therefore $\det \nabla T \geq 0$ and the transport condition $|\det \nabla T| = 1$ implies that

$$1 = \det \nabla T = \prod_{k=1}^n \lambda_k(x), \quad \lambda_k \geq 0 \qquad \text{on } E.$$

By the arithmetic geometric mean inequality,

$$1 = (\det \nabla T)^{1/n} \leq \frac{\text{div } T}{n}, \tag{2.15}$$

therefore we can repeat Gromov's argument and prove the Wulff inequality

$$n|K|^{1/n}|E|^{(n-1)/n} = n|E| = n \int_E (\det \nabla T)^{1/n}$$
$$\leq \int_E \text{div } T = \int_{\partial E} T \cdot \nu_E \leq \int_{\partial E} f_*(T) \, f(\nu_E) \leq \mathcal{F}(E),$$

as $T(E) \subset K$ and $K = \{f_* < 1\}$. The characterization of equality cases is also immediate: equality holding in the arithmetic geometric mean inequality (2.15) we have again that, a.e. on E, $\lambda_k = 1$ for every $1 \leq k \leq n$. The more rigid structure of the *optimal* transport map is then exploited through (2.14), from which we deduce

$$\nabla T = \text{Id} \quad \text{on } E.$$

Provided E is connected (see Remark 2.2) this implies the existence of $x_0 \in \mathbb{R}^n$ such that $T(x) = x - x_0$ on E, and therefore $E = x_0 + K$, as required.

Remark 2.2. The connectedness of minimizers in the Wulff inequality can eventually be justified as follows. If E is optimal in the Wulff inequality and admits a decomposition $E = E_1 \cup E_2$ for E_1, E_2 such that $\mathcal{F}(E) = \mathcal{F}(E_1) + \mathcal{F}(E_2)$ and $|E| = |E_1| + |E_2|$, then

$$n|K|^{1/n}|E|^{(n-1)/n} = \mathcal{F}(E) = \mathcal{F}(E_1) + \mathcal{F}(E_2)$$
$$\geq n|K|^{1/n}(|E_1|^{(n-1)/n} + |E_2|^{(n-1)/n}).$$

Therefore, by the strict concavity of $t \mapsto t^{(n-1)/n}$, we find that

$$\min\{|E_1|, |E_2|\} = 0.$$

3 Symmetry and stability

We now start our discussion of the stability properties of the Euclidean isoperimetric inequality: if the perimeter of a set E is close to the perimeter of a ball having the same measure, is the set E close to be a ball? To this purpose we introduce the *isoperimetric deficit* of E, defined as

$$\delta(E) = \frac{P(E)}{n|B|^{1/n}|E|^{(n-1)/n}} - 1.$$

By the Euclidean isoperimetric inequality, we always have $\delta(E) \geq 0$, with $\delta(E) = 0$ if and only if E is a ball. The isoperimetric deficit is a scale and translation invariant set function, *i.e.* $\delta(E) = \delta(x + r E)$ for every $r > 0$ and $x \in \mathbb{R}^n$. The stability problem thus amounts in studying sets E such that $\delta(E)$ is small, trying to express quantitative, explicit estimates on their distance from the set of balls of $\mathbb{R}^n$.

The choice of the distance is of course crucial. For example, a set may have arbitrarily small isoperimetric deficit and, at the same time, arbitrarily large Hausdorff distance from the set of balls (see Figure 3.1). In dimension $n = 2$ this can be realized on adding, far away from a "main ball", an arbitrary number of connected components of small total perimeter. In dimension $n \geq 3$ this kind of construction can be performed without even losing connectedness, by attaching to a "main ball" some long, tiny tentacles of small total perimeter. These constructions prevent the possibility to control the Hausdorff distance from the set of balls in terms of the isoperimetric deficit. Note however that, in both examples, the Lebesgue measure of the symmetric difference between the sets we have constructed and their "main ball" is small, and has to decrease to zero if we make the isoperimetric deficit vanish.

The above considerations suggest that, for a stability estimate to hold true on generic sets, we should consider the set function

$$\alpha(E) = \inf_{x \in \mathbb{R}^n} \left\{ \frac{|E \Delta B(x,r)|}{|E|} : |B(x,r)| = |E| \right\},$$

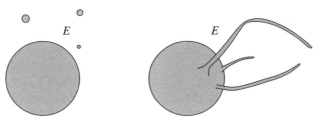

Figure 3.1. Sets with small isoperimetric deficit and large Hausdorff distance from the set of balls.

called the *Fraenkel asymmetry* of E. Just like the isoperimetric deficit, the Fraenkel asymmetry is a scale and translation invariant set function that vanishes on the set of balls. The following theorem provides a stability estimate for the Euclidean isoperimetric inequality.

Theorem 3.1. *There exists a constant $C(n)$ depending on the dimension n only such that*

$$\delta(E) \geq C(n)\alpha(E)^2. \tag{3.1}$$

Remark 3.2. We are interested in an inequality like (3.1) in the *small deficit regime*

$$\delta(E) \leq \delta(n), \tag{3.2}$$

where $\delta(n) > 0$ is some given constant. Indeed we always have $\alpha(E) \leq 2$, so that (3.1) trivially holds with $C(n) \geq 4\delta(n)^{-1}$ whenever $\delta(E) \geq \delta(n)$.

Remark 3.3. Combining the characterization of equality cases in the Euclidean isoperimetric inequality with some standard compactness arguments from the theory of sets of finite perimeter, it is not hard to prove that for every $\varepsilon > 0$ there exists $\delta > 0$ such that $\delta(E) \leq \delta$ implies $\alpha(E) \leq \varepsilon$. Therefore the interest of (3.1) is in that it provides an explicit (quadratic) decay rate of the Fraenkel asymmetry in terms of the isoperimetric deficit. This decay rate is optimal in the sense that we cannot strengthen (3.1) into $\delta(E) \geq C(n)\alpha(E)^p$ for some $p \in (0, 2)$. Indeed, it is easy to construct a sequence of sets $\{E_h\}_{h \in \mathbb{N}}$ such that

$$\lim_{h \to \infty} \delta(E_h) = 0, \qquad \limsup_{h \to \infty} \frac{P(E_h)}{\alpha(E_h)^2} < \infty,$$

by looking at a sequence of ellipsoids converging to B (for another example see [40, Figure 4]).

Theorem 3.1, first proved in the above generality in [28], has a long history. On connected sets in the plane it was obtained by Bernstein [2] and Bonnesen [6] at the beginning of the past century. The first results in higher dimension are more recent. Fuglede [25] proves Theorem 3.1 in the class of convex sets. In this case one can actually aim to control the Hausdorff distance from the set of balls, and this was done (again with the sharp decay rates) in [26]. Hall [35] has proved Theorem 3.1 in the class of axially symmetric sets. A common treat of all these results is the existence of a global parametrization for the boundary of the set E under consideration. Convex sets (or nearly spherical star-shaped domains) can be naturally parameterized by scalar functions defined on a reference sphere, and the perimeter of a planar connected set is decreased by convexification. In the same vein, axially symmetric sets are completely described in terms of functions of one real variable. Starting from Hall's result for axially symmetric sets, a non-sharp estimate on generic sets

$$\delta(E) \geq C(n)\alpha(E)^4,$$

was obtained by Hall, Hayman and Weitsmann in [36]. Their idea is to reduce to the case of axially symmetric sets by showing that for every set E there exists an axially symmetric set F such that

$$\alpha(E)^2 \leq C(n)\alpha(F), \qquad \delta(F) \leq \delta(E). \tag{3.3}$$

Then, by Hall's result one easily obtains

$$\alpha(E)^4 \leq C(n)\alpha(F)^2 \leq C(n)\delta(F) \leq C(n)\delta(E),$$

as claimed. In order to construct a set F such that (3.3) holds true, they rely on symmetrization theory, in particular on the notion of Schwartz symmetrization. Given a set E and a direction v, the *Schwartz symmetrization* of E with respect to v is the set E_v^* whose slice

$$E_v^* \cap \{x \in \mathbb{R}^n : x \cdot v = t\}, \quad t \in \mathbb{R},$$

is given by the $(n-1)$-dimensional disk orthogonal to v, centered at $t\,v$, and having the same $(n-1)$-dimensional measure as $E \cap \{x \in \mathbb{R}^n : x \cdot v = t\}$, see Figure 3.2 (and (3.11) for the rigorous definition). By construction, E_v^* is axially symmetric and $|E_v^*| = |E|$. Moreover, as a consequence of the isoperimetric inequality itself, one can show the *Schwartz inequality*

$$P(E_v^*) \leq P(E), \tag{3.4}$$

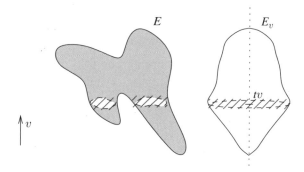

Figure 3.2. The Schwarz symmetrization E_v of E with respect to the direction v.

Figure 3.3. n this case $\alpha(E) > 0$ but $\alpha(E_v^*) = 0$ for an unlucky choice of v.

that in particular implies

$$\delta(E_v^*) \leq \delta(E). \tag{3.5}$$

Thus the second inequality in (3.3) is actually true with $F = E_v^*$ for *every* $v \in S^{n-1}$. On the contrary, depending on the choice of v and E, it may happen that $\alpha(E) > 0$ but $\alpha(E_v^*) = 0$, see Figure 3.3. What is shown in [36] is the existence of a constant $C(n)$ such that every set E admits a suitable symmetrization direction v with the property that

$$\frac{\alpha(E)^2}{\alpha(E_v^*)} \leq C(n). \tag{3.6}$$

We note that to obtain the sharp inequality (3.1) with the same strategy one would need to control the ratio $\alpha(E)/\alpha(E_v^*)$ but, as shown in [36], this is actually impossible since (3.6) is optimal on generic sets.

We now proceed as follows. In Sections 3.1-3.2 we describe the proof of Theorem 3.1 from [28]. The idea of relying on reduction inequalities like (3.3) is still present, introducing however an intermediate symmetrization step between general sets and axially symmetric sets, in order to preserve sharp decay rates in the chain of reduction inequalities. Before coming to this, we conclude this section with a sketch of the proof of (3.4).

Sketch of the proof of the Schwarz inequality.

Step 1. If M is a $(n-1)$-dimensional manifold of class C^1 and $v_M : M \rightarrow S^{n-1}$ is a Borel measurable vector field with the property that $T_x M = \langle v_M(x) \rangle^\perp$, then we have the following Fubini type theorem on M

$$\int_{\mathbb{R}} \mathcal{H}^{n-2}(M \cap \{x_1 = t\})dt = \int_M \sqrt{1 - (v_M \cdot e_1)^2}d\mathcal{H}^{n-1}. \tag{3.7}$$

Indeed, when M is contained in some $(n-1)$-dimensional plane in $\mathbb{R}^n$ this is just a consequence of Pythagoras' theorem. The general case follows by decomposing M into a finite number of pieces, each one of them admitting an almost-flat parametrization. Starting from (3.7) and by a standard approximation argument we see that

$$\int_{\mathbb{R}} dt \int_{M \cap \{x_1 = t\}} g(x) d\mathcal{H}^{n-2}(x) = \int_M g \sqrt{1 - (\nu_M \cdot e_1)^2} d\mathcal{H}^{n-1}, \quad (3.8)$$

for every Borel function $g : M \to [0, \infty]$. Let us now consider the Borel set

$$M_0 = \{x \in M : |\nu_M(x) \cdot e_1| = 1\},$$

and define the Borel function $g : M \to [0, \infty]$ on setting

$$g(x) = \frac{1}{\sqrt{1 - (\nu_M(x) \cdot e_1)^2}}, \quad x \in M \setminus M_0,$$

and $g(x) = +\infty$ if $x \in M_0$. Plugging this choice of g into (3.8) we find that, if $\mathcal{H}^{n-1}(M_0) = 0$, then

$$\mathcal{H}^{n-1}(M) = \int_{\mathbb{R}} dt \int_{M \cap \{x_1 = t\}} \frac{d\mathcal{H}^{n-2}}{\sqrt{1 - (\nu_M \cdot e_1)^2}}. \quad (3.9)$$

In particular, provided $\mathcal{H}^{n-1}(M_0) = 0$,

$$\int_M f \, d\mathcal{H}^{n-1} = \int_{\mathbb{R}} dt \int_{M \cap \{x_1 = t\}} \frac{f \, d\mathcal{H}^{n-2}}{\sqrt{1 - (\nu_M \cdot e_1)^2}}, \quad (3.10)$$

for every Borel function $f : M \to [0, \infty]$.

Step 2. Let us now consider the Schwarz symmetrization $E^* = E_\nu^*$ of E relative to the choice $\nu = e_1$, i.e.

$$E^* = \left\{ x \in \mathbb{R}^n : |x'| < \left(\frac{v(x_1)}{\omega_{n-1}} \right)^{1/(n-1)} \right\}, \quad (3.11)$$

where we have set

$$v(t) = \mathcal{H}^{n-1}(E \cap \{x_1 = t\}), \quad t \in \mathbb{R},$$

and where $x = (x_1, x') \in \mathbb{R} \times \mathbb{R}^{n-1}$ for $x \in \mathbb{R}^n$. Note that, by construction,

$$\mathcal{H}^{n-1}(E^* \cap \{x_1 = t\}) = v(t), \quad \forall t \in \mathbb{R}.$$

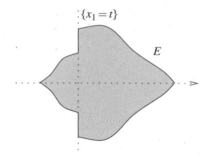

Figure 3.4. The continuity of v is related to the validity of (3.12).

We claim that, provided

$$\mathcal{H}^{n-1}(\{x \in \partial E : |\nu_E(x) \cdot e_1| = 1\}) = 0, \qquad (3.12)$$

then v is absolutely continuous on $\mathbb{R}$ with distributional derivative given by

$$v'(t) = -\int_{\{x_1=t\}\cap\partial E} \frac{(\nu_E \cdot e_1)}{\sqrt{1-(\nu_E \cdot e_1)^2}}\, d\mathcal{H}^{n-2}, \qquad (3.13)$$

see Figure 3.4. Indeed let $\varphi \in C_c^1(\mathbb{R})$ and set $u(x) = \varphi(x_1)$. Then by Fubini's theorem, by the divergence theorem and by (3.10) we have

$$\int_{\mathbb{R}} \varphi'(t)v(t)dt = \int_{\mathbb{R}} dt \int_{E\cap\{x_1=t\}} \frac{\partial u}{\partial x_1}(x)d\mathcal{H}^{n-1}(x)$$

$$= \int_E \frac{\partial u}{\partial x_1} = \int_{\partial E}(\nu_E \cdot e_1)\, u\, d\mathcal{H}^{n-1}$$

$$= \int_{\mathbb{R}} \varphi(t)dt \int_{\{x_1=t\}\cap\partial E} \frac{(\nu_E \cdot e_1)}{\sqrt{1-(\nu_E \cdot e_1)^2}}d\mathcal{H}^{n-2},$$

as required.

Step 3. We now prove the Schwarz inequality (3.4). For every $\varepsilon > 0$ we can find an orthogonal map $Q_\varepsilon \in \mathbf{SO}(n)$ such that $|Q_\varepsilon - \mathrm{Id}_{\mathbb{R}^n}| < \varepsilon$ and $Q_\varepsilon(E)$ satisfies (3.12). Since $P(Q_\varepsilon(E)) = P(E)$ and $P(Q_\varepsilon(E^*)) = P(E^*)$, we can directly assume that (3.12), and thus (3.13), hold true for E. Let us now set

$$p(t) = \mathcal{H}^{n-2}(\{x_1 = t\} \cap \partial E), \quad p_*(t) = \mathcal{H}^{n-2}(\{x_1 = t\} \cap \partial E^*).$$

By (3.9) we have that

$$P(E) = \mathcal{H}^{n-1}(\partial E) = \int_{\mathbb{R}} dt \int_{\{x_1=t\}\cap\partial E} \frac{d\mathcal{H}^{n-2}}{\sqrt{1-(\nu_E\cdot e_1)^2}}$$

$$= \int_{\mathbb{R}} p(t)dt \fint_{\{x_1=t\}\cap\partial E} \sqrt{1 + \frac{(\nu_E\cdot e_1)^2}{1-(\nu_E\cdot e_1)^2}}\, d\mathcal{H}^{n-2}$$

$$\geq \int_{\mathbb{R}} p(t)\sqrt{1 + \left(\fint_{\{x_1=t\}\cap\partial E} \frac{(\nu_E\cdot e_1)}{\sqrt{1-(\nu_E\cdot e_1)^2}}\right)^2}\, dt \tag{3.14}$$

$$= \int_{\mathbb{R}} \sqrt{p(t)^2 + v'(t)^2}\, dt,$$

where we have applied the Jensen's inequality to $z \mapsto \sqrt{1+z^2}$ and (3.13). In case we repeat this argument with E^* in place of E we clearly have equality in the application of the Jensen's inequality, as $\nu_{E^*}\cdot e_1$ is constant over $\{x_1 = t\}\cap\partial E$. Therefore we also have

$$P(E^*) = \int_{\mathbb{R}} \sqrt{p_*(t)^2 + v'(t)^2}\, dt. \tag{3.15}$$

By the Euclidean isoperimetric inequality (in $\mathbb{R}^{n-1}$), we have $p(t) \geq p_*(t)$, therefore $P(E) \geq P(E^*)$, as required. □

Remark 3.4. The argument sketched above shows in fact that

$$P(E; S) \geq P(E^*; S), \tag{3.16}$$

whenever S is a stripe of the form $S = I \times \mathbb{R}^{n-1}$, for an open set $I \subset \mathbb{R}^n$. Moreover we easily see that if $P(E; S) = P(E^*; S)$ and (3.12) holds true then $p(t) = p_*(t)$ for a.e. $t \in \mathbb{R}$, so that the slices $\{x_1 = t\}\cap E$ of E are $(n-1)$-dimensional balls, and moreover $\nu_E\cdot e_1$ is constant along $\{x_1 = t\}\cap\partial E$. Note that if this last condition fails then $P(E) > P(E^*)$ even if all the slices $\{x_1 = t\}\cap E$ are $(n-1)$-dimensional balls, see Figure 3.5.

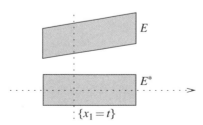

Figure 3.5. If $\nu_E\cdot e_1$ is not constant along $\{x_1 = t\}\cap\partial E$ then $P(E) > P(E^*)$ even if all the slices $\{x_1 = t\}\cap E$ are $(n-1)$-dimensional balls.

3.1 From generic sets to n-symmetric sets

A set $E \subset \mathbb{R}^n$ is *n-symmetric* provided E is symmetric by reflection with respect to n mutually orthogonal hyperplanes. The intersection of these hyperplanes is called the center of symmetry of E. The optimal center for determining the value of $\alpha(E)$ is not, in general, the center of symmetry of E (see Figure 3.6).

It is however true that the distance of E from the ball (having the measure $|E|$) and centered at its center of symmetry is, at most, three times its Fraenkel asymmetry, *i.e.*

$$\alpha(E) \leq \frac{|E \Delta B|}{|E|} \leq 3\alpha(E), \qquad (3.17)$$

where we are assuming that $|E| = |B|$ and that the center of symmetry of E lies at the origin. The first inequality in (3.17) is trivial by the definition of α. To prove the second inequality, let $x \in \mathbb{R}^n$ be such that $\alpha(E) = |E \Delta B(x)|$, $B(x) = x + B$. Since E is n-symmetric and it has center at the origin we have that $E = \{-x : x \in E\}$. Therefore $|E \Delta B(-x)| = |E \Delta B(x)| = \alpha(E)$. Since $|B \Delta B(x)| \leq |B(-x) \Delta B(x)|$ we thus find that

$$
\begin{aligned}
|E \Delta B| &\leq |E \Delta B(x)| + |B(x) \Delta B| \\
&\leq |E \Delta B(x)| + |B(x) \Delta B(-x)| \\
&\leq |E \Delta B(x)| + |B(x) \Delta E| + |E \Delta B(-x)| = 3\,\alpha(E),
\end{aligned}
$$

as required. The interesting consequence of (3.17) is that, when proving (3.1) on a n-symmetric set, we can replace the Fraenkel asymmetry $\alpha(E)$ of E with the distance of E from the ball centered in its center of symmetry. Since E and its center of symmetry are connected by a strong symmetry property this fact reveals particularly useful, as we are going

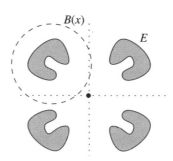

Figure 3.6. The ball $B(x)$ satisfies $\alpha(E) = |E \Delta B(x)|$, however $|E \Delta B| \leq 3|E \Delta B(x)|$.

to see in many occasions. We now discuss the reduction theorem that allows to pass from generic sets to n-symmetric sets when proving the quantitative Euclidean isoperimetric inequality.

Theorem 3.5. *For every $E \subset \mathbb{R}^n$ there exists a n-symmetric set F such that*

$$\alpha(E) \leq C(n)\alpha(F), \quad \delta(F) \leq C(n)\delta(E). \quad (3.18)$$

Remark 3.6. We are going to prove a more precise result. Given a set $E \subset \mathbb{R}^n$ and an orthonormal basis $\{v_h\}_{h=1}^n$ of $\mathbb{R}^n$, up to possibly change some v_h with $-v_h$, we shall find a set of the form

$$G = \{x \in E : x \cdot v_h > \alpha_h\}, \quad \alpha_h \in \mathbb{R},$$

with $|G| = 2^{-n}|E|$ and such that, by iteratively reflecting G with respect to the hyperplanes $\{x \cdot v_h = \alpha_h\}$, we may construct a set F satisfying (3.18).

Sketch of the proof of Theorem 3.5. Without loss of generality we may assume that $|E| = |B|$. Arguing as in Remark 3.2 we reduce to consider the case $\delta(E) \leq \delta(n)$, for a given positive constant $\delta(n)$.

Step 1. We prove (3.18) with F having just one hyperplane of symmetry. Given $v \in S^{n-1}$ we consider a half-space H^+ of the form

$$H_v^+ = \{x \in \mathbb{R}^n : x \cdot v > t\}$$

such that $|E \cap H_v^+| = |E|/2$, and let $H_v^- = \mathbb{R}^n \setminus H^+$. If $\pi : \mathbb{R}^n \to \mathbb{R}^n$ is the reflection with respect to the hyperplane $H_v = \{x \in \mathbb{R}^n : x \cdot v = t\}$, then we set, see Figure 3.7,

$$E_v^+ = (E \cap H^+) \cup \pi(E \cap H^+), \qquad E_v^- = (E \cap H^-) \cup \pi(E \cap H^-).$$

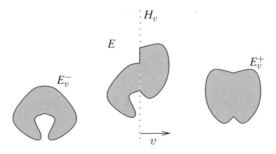

Figure 3.7. The construction of E_v^+ and E_v^-.

By construction both E_ν^+ and E_ν^- have a hyperplane of symmetry. We also have

$$|E_\nu^+| = |E_\nu^-| = |E| = |B|. \tag{3.19}$$

Moreover the perimeters of E_ν^+ and E_ν^- satisfy

$$P(E_\nu^+) = 2\mathcal{H}^{n-1}(H_\nu^+ \cap \partial E), \quad P(E_\nu^-) = 2\mathcal{H}^{n-1}(H_\nu^- \cap \partial E)$$

while

$$P(E) = \mathcal{H}^{n-1}(H_\nu^+ \cap \partial E) + \mathcal{H}^{n-1}(H_\nu^- \cap \partial E) + \mathcal{H}^{n-1}(\partial E \cap H_\nu),$$

so that

$$\frac{P(E_\nu^+) + P(E_\nu^-)}{2} \leq P(E). \tag{3.20}$$

Hence we combine (3.19) and (3.20) to find that

$$\frac{\delta(E_\nu^+) + \delta(E_\nu^-)}{2} \leq \delta(E). \tag{3.21}$$

In particular, $\max\{\delta(E_\nu^+), \delta(E^-\nu)\} \leq 2\,\delta(E)$ and the second inequality in (3.18) is satisfied by both E_ν^+ and E_ν^-. Let now B_ν^+ be a ball centered on H_ν such that

$$|E_\nu^+ \Delta B_\nu^+| \leq |E_\nu^+ \Delta(x + B)|, \quad \forall x \in H_\nu,$$

i.e. B_ν^+ is a closest ball to E_ν^+ among those centered on H_ν, and let B_ν^- be defined likewise on starting from E_ν^-. Then we have the following basic estimate,

$$|B|\alpha(E) \leq |E\Delta B_\nu^+| = \frac{|E_\nu^+ \Delta B_\nu^+| + |E_\nu^- \Delta B_\nu^+|}{2}$$

$$\leq \frac{|E_\nu^+ \Delta B_\nu^+| + |E_\nu^- \Delta B_\nu^-| + |B_\nu^+ \Delta B_\nu^-|}{2}.$$

By an argument analogous to that leading to (3.17) we see that, since E_ν^+ is symmetric by reflection with respect to H_ν, then

$$\frac{|E_\nu^+ \Delta B_\nu^+|}{|E_\nu^+|} \leq 3\alpha(E_\nu^+).$$

Since a similar relation holds for E_ν^- we thus find that

$$\alpha(E) \leq C(\alpha(E_\nu^+) + \alpha(E_\nu^-) + |B_\nu^+ \Delta B_\nu^-|), \tag{3.22}$$

for a constant C. This estimate is essentially optimal. Indeed if we think to the situation in Figure 3.3, we see that $\alpha(E)$ could be positive and comparable with $|B_\nu^+ \Delta B_\nu^-|$, while $\alpha(E_\nu^+) = \alpha(E_\nu^-) = 0$. Therefore if we just pick a direction $\nu \in S^{n-1}$ and chose $F \in \{E_\nu^+, E_\nu^-\}$ we cannot expect (3.18) to hold true. The idea is now to use the assumption $\delta(E) \le \delta(n)$ to prove that if a direction ν_1 is so unlucky that $\alpha(E_1^+), \alpha(E_1^-) \approx 0$ while $|B_1^+ \Delta B_1^-| \approx \alpha(E)$ then every direction ν_2 that is orthogonal to ν_1 is such that $\alpha(E) \approx \max\{\alpha(E_2^+), \alpha(E_2^-)\}$. For the details, we refer the reader to [28, Section 2] or [40, Section 6].

Step 2. Let $\{\nu_k\}_{k=1}^n$ be an orthonormal basis of $\mathbb{R}^n$. A simple iteration of step one shows that there exists a set F that is symmetric by reflection with respect to $(n-1)$-mutually orthogonal hyperplanes (that, up to a rotation, we can assume to be oriented by the vectors $\{\nu_k\}_{k=1}^{n-1}$), and such that

$$\alpha(E) \le C(n)\alpha(F), \quad \delta(F) \le 2^{n-1}\delta(E). \quad (3.23)$$

We cannot iterate further the argument of step one. Indeed, if we want to preserve the $(n-1)$-symmetries gathered so far, we are forced to symmetrize by reflection F with respect to the direction ν_n: but in this way the possibility of choosing among two orthogonal directions is precluded. However, since F is $(n-1)$-symmetric, if we perform the construction of step one with respect to $\nu = \nu_n$ we are led to chose the balls B_ν^+ and B_ν^- so that they are centered at the *common* center of symmetry of the n-symmetric sets F_ν^+ and F_ν^-, i.e. we actually have $B_\nu^+ = B_\nu^-$. Therefore (3.22) gives

$$\alpha(F) \le C(\alpha(F_\nu^+) + \alpha(F_\nu^-)).$$

Since $\max\{\delta(F_\nu^+), \delta(F_\nu^-)\} \le 2\delta(F)$, by (3.23) we obtain

$$\alpha(E) \le C(n)\alpha(G), \quad \delta(G) \le 2^n \delta(E),$$

either for $G = F_\nu^+$ or for $G = F_\nu^-$. $\square$

3.2 Sketch of the proof of Theorem 3.1

Step 1. By rotating E we can assume without loss of generality that

$$\mathcal{H}^{n-1}(\{x \in \partial E : |\nu_E(x) \cdot e_i| = 1\}) = 0, \quad (3.24)$$

for every $i = 1, \dots, n$. By Theorem 3.5 and by Remark 3.6 we can find a set F that is symmetric with respect to the coordinate directions and such that

$$\alpha(E) \le C(n)\alpha(F), \quad \delta(F) \le C(n)\delta(E).$$

Since F is obtained by iteratively reflecting a portion of E, see again Remark 3.6, by (3.24) we have that

$$\mathcal{H}^{n-1}(\{x \in \partial F : |v_F(x) \cdot e_i| = 1\}) = 0,$$

for every $i = 1, \ldots, n$. In conclusion we can directly assume that E satisfies (3.24) and that it is symmetric with respect to the coordinate directions. We can further assume that $|E| = |B|$. To prove Theorem 3.1 we are now going to show that

$$|E \Delta B| \leq C(n)\sqrt{\delta(E)}. \tag{3.25}$$

Let us first see that (3.25) can be reduced to prove that E is close to its Schwarz symmetrization with respect to one of the coordinate axes, see (3.28). To show this, let us consider the two stripes

$$S_k = \left\{ x \in \mathbb{R}^n : |x_k| < \frac{1}{\sqrt{2}} \right\}, \quad k = 1, 2.$$

Since $B \subset S_1 \cup S_2$ we have that

$$\max\{|(B \setminus E) \cap S_1|, |(B \setminus E) \cap S_2|\} \geq \frac{|B \setminus E|}{2} = \frac{|E \Delta B|}{4}.$$

Up to a rotation we may assume that

$$|E \Delta B| \leq 4|(B \setminus E) \cap S_1| \leq 4(|(E^* \setminus E) \cap S_1| + |(B \setminus E^*) \cap S_1|), \tag{3.26}$$

where we have introduced the Schwarz symmetrization E^* of E with respect to the direction x_1. Let us now notice that E^* is symmetric by reflection with respect to the coordinate axes. Thus by (3.17) we have that

$$|E^* \Delta B| \leq 3|B|\alpha(E^*) \leq C(n)\sqrt{\delta(E^*)} \leq C(n)\sqrt{\delta(E)}, \tag{3.27}$$

where we have also applied Hall's theorem [35] (*i.e.*, Theorem 3.1 on axially symmetric sets – see [28, Section 4] and [40, Section 7.2] for alternative proofs) and the Schwarz inequality (3.5). Thus, combining (3.26) and (3.27), we are left to show that

$$|(E \Delta E^*) \cap S_1| \leq C(n)\sqrt{\delta(E)}, \tag{3.28}$$

in order to complete the proof of (3.25).

Step 2. Let us define

$$v(t) = \mathcal{H}^{n-1}(E \cap \{x_1 = t\}), \quad w(t) = \mathcal{H}^{n-1}(B \cap \{x_1 = t\}).$$

By Remark 3.3 we know that for every $\varepsilon > 0$ there exists $\delta > 0$ such that $\alpha(E) \leq \varepsilon$ if $\delta(E) \leq \delta$. At the same time w is bounded from above on $\mathbb{R}$ and it is strictly positive on $[-1/\sqrt{2}, 1/\sqrt{2}]$. Since $\int_{\mathbb{R}} |v - w| = |E \Delta B| \leq \alpha(E)$, starting from these remarks it is possible to prove that

$$v(t) \leq C(n), \quad t \in \mathbb{R}, \tag{3.29}$$

$$v(t) \geq c(n), \quad |t| \leq \frac{1}{\sqrt{2}}, \tag{3.30}$$

provided $\delta(E) \leq \delta(n)$ for $\delta(n)$ small enough. By the Schwarz inequality

$$
\begin{aligned}
C(n)\delta(E) \quad = \quad & P(E) - P(B) \geq P(E) - P(E^*) \\
& + \underbrace{P(E^*) - P(B)}_{\geq 0 \text{ by } (1.1)} \geq P(E) - P(E^*) \\[2mm]
\overset{(3.24)}{=\!=} \quad & P(E; S_1) - P(E^*; S_1) \\
& + \underbrace{P(E; \mathbb{R}^n \setminus S_1) - P(E^*; \mathbb{R}^n \setminus S_1)}_{\geq 0 \text{ by } (3.16)} \\[2mm]
\overset{(3.14) \text{ and } (3.24)}{\geq} \quad & \int_I \sqrt{p(t)^2 + v'(t)^2}\,dt - P(E^*; S_1) \\[2mm]
\overset{(3.15)}{=\!=} \quad & \int_I \sqrt{p(t)^2 + v'(t)^2} - \sqrt{p_*(t)^2 + v'(t)^2}\,dt,
\end{aligned}
$$

where $I = (-1/\sqrt{2}, 1/\sqrt{2})$,

$$p(t) = \mathcal{H}^{n-1}(\{x_1 = t\} \cap \partial E), \qquad p_*(t) = \mathcal{H}^{n-1}(\{x_1 = t\} \cap \partial E^*).$$

We now remark that

$$\sqrt{p(t)^2 + v'(t)^2} - \sqrt{p_*(t)^2 + v'(t)^2}$$

$$= \frac{p(t)^2 - p_*(t)^2}{\sqrt{p(t)^2 + v'(t)^2} + \sqrt{p_*(t)^2 + v'(t)^2}},$$

therefore by Hölder's inequality

$$\int_I \sqrt{p(t)^2 - p_*(t)^2}\,dt$$

$$\leq \sqrt{\int_I \sqrt{p(t)^2 + v'(t)^2} + \sqrt{p_*(t)^2 + v'(t)^2}\,dt}\sqrt{C(n)\delta(E)}$$

$$\leq C(n)\sqrt{P(E) + P(E^*)}\sqrt{\delta(E)} \leq C(n)\sqrt{\delta(E)},$$

as $P(E^*) \leq P(E) \leq P(B)(1 + \delta(E))$. By the isoperimetric inequality in dimension $(n - 1)$ we have $p(t) \geq p_*(t)$, therefore

$$C(n)\sqrt{\delta(E)} \geq \int_I \sqrt{(p(t) - p_*(t))(2p_*(t))} = \sqrt{2} \int_I p_*(t)\sqrt{\frac{p(t)}{p_*(t)} - 1}\, dt.$$

Since $p_*(t) = c(n)v(t)^{(n-2)/(n-1)}$, by (3.30) we find that

$$C(n)\sqrt{\delta(E)} \geq \int_I \sqrt{\frac{p(t)}{p_*(t)} - 1}\, dt. \qquad (3.31)$$

We now notice that

$$\sqrt{\frac{p(t)}{p_*(t)} - 1},$$

is the isoperimetric deficit (in $\mathbb{R}^{n-1}$) of the $(n - 1)$-dimensional set $E_t = \{x_1 = t\} \cap E$. On arguing by induction (the case $n = 2$ being trivial from this argument), we have that

$$\sqrt{\frac{p(t)}{p_*(t)} - 1} \geq \text{Fraenkel asymmetry of } E_t \quad \text{in } \mathbb{R}^{n-1}.$$

Since E_t is symmetric with respect to the $(n-1)$ coordinate directions of $\mathbb{R}^{n-1}$ and since $E_t^* = E^* \cap \{x_1 = t\}$ is a $(n-1)$-dimensional ball centered at the origin of $\mathbb{R}^{n-1}$ with $\mathcal{H}^{n-1}(E_t^*) = v(t) = \mathcal{H}^{n-1}(E_t)$, we thus have

$$\sqrt{\frac{p(t)}{p_*(t)} - 1} \geq \frac{1}{3}\frac{\mathcal{H}^{n-1}(E_t \Delta E_t^*)}{\mathcal{H}^{n-1}(E_t)}.$$

Thus by (3.29) and by (3.31) we find that

$$|(E \Delta E^*) \cap S_1| \leq \int_I \mathcal{H}^{n-1}(E_t \Delta E_t^*)\, dt$$

$$\leq 3 \int_I v(t)\sqrt{\frac{p(t)}{p_*(t)} - 1}\, dt \leq C(n)\sqrt{\delta(E)}.$$

This proves (3.28) and thus concludes the proof of Theorem 3.1.

4 Mass transport and stability

Of course one cannot hope to attack the stability problem for the Wulff inequality by the symmetrization methods developed in the proof of Theorem 3.1. Indeed, these methods rest on the full symmetry of B, a property that is clearly missing in the case of a generic Wulff shape K. An entirely different approach to the stability problem, based on optimal mass transportation theory and Gromov's proof of the Wulff inequality, has been recently developed in [21], and it leads to the following result. Let us mention that a non-sharp decay rate was first obtained in [20] on moving from the characterization of equality cases for the Wulff inequality provided in [24].

Theorem 4.1. *There exists a constant $C(n)$ depending on the dimension n only, such that*

$$\delta(E; K) \geq C(n)\alpha(E; K)^2, \tag{4.1}$$

where we have set

$$\delta(E; K) = \frac{\mathcal{F}(E)}{n|K|^{1/n}|E|^{(n-1)/n}} - 1,$$

$$\alpha(E; K) = \inf_{x \in \mathbb{R}^n} \left\{ \frac{|E \Delta (x + rK)|}{|E|} : r^n |K| = |E| \right\}.$$

Remark 4.2. Theorem 4.1 covers Theorem 3.1 as a particular case. Moreover the constant $C(n)$ in (4.1) is polynomial in the dimension, *i.e.* $C(n) \approx n^7$, while the reduction step to n-symmetric sets performed in Theorem 3.5 forces $C(n) \approx 2^n$ in (3.1).

Remark 4.3. We stress the fact that the constant $C(n)$ appearing in (4.1) is independent of K. More precisely, we shall first prove the inequality

$$\delta(E; K) \geq C(n, K)\alpha(E; K)^2, \tag{4.2}$$

where $C(n, K)$ depends on the dimension n and on the ratio M_K/m_K only, for

$$m_K = \inf\{f(v) : v \in S^{n-1}\}, \quad M_K = \sup\{f(v) : v \in S^{n-1}\}.$$

Then one notices that by John Lemma there exists an affine map $T_0 : \mathbb{R}^n \to \mathbb{R}^n$ such that

$$\det T_0 > 0, \quad B \subset T_0(K) \subset B_n = \{x : |x| < n\}.$$

In particular, there exists an affine map $T : \mathbb{R}^n \to \mathbb{R}^n$ such that

$$\det T = 1, \quad B_r \subset T(K) \subset B_{rn},$$

for some $r > 0$. We notice that $T(K)$ is an open, bounded convex set containing the origin, with $M_{T(K)}(m_{T(K)})^{-1} \leq n$. If $\mathcal{F}_{T(K)}$ is the anisotropic surface energy associated with $T(K)$ as in Remark 2.1, then by (4.2) we have that

$$C(n)\sqrt{\delta(T(E); T(K))} \geq \alpha(T(E); T(K)).$$

Since $\det T = 1$ we easily find that $\alpha(T(E); T(K)) = \alpha(E; K)$ and that

$$\mathcal{F}_{T(K)}(T(E)) = \lim_{\varepsilon \to 0^+} \frac{|T(E) + \varepsilon T(K)| - |T(E)|}{\varepsilon}$$

$$= \lim_{\varepsilon \to 0^+} \frac{|E + \varepsilon K| - |E|}{\varepsilon} = \mathcal{F}(E),$$

where we have used (2.4). Thus $\delta(T(E); T(K)) = \delta(E; K)$ and we are done.

We now introduce the problem to turn Gromov's proof of the Wulff inequality into a quantitative estimate like (4.1). In order to simplify the notation we shall directly consider the Euclidean case

$$K = B$$

where $f(v) = f_*(v) = 1$ for every $v \in S^{n-1}$. We correspondingly set

$$\alpha(E; B) = \alpha(E), \quad \delta(E; B) = \delta(E).$$

We let $|E| = |B|$ and denote by T the Brenier map between E and B, so that

$$\nabla T = \sum_{k=1}^n \lambda_k \, e_k \otimes e_k,$$

where the λ_k's are positive, with $\lambda_k \leq \lambda_{k+1}$, and such that

$$1 = \det \nabla T = \prod_{k=1}^n \lambda_k \quad \text{on } E.$$

If we denote the geometric mean and the arithmetic mean of the λ_k's with

$$\lambda_G = \prod_{k=1}^n \lambda_k^{1/n}, \quad \lambda_A = \frac{1}{n} \sum_{k=1}^n \lambda_k,$$

then by Gromov's argument we deduce the following lower bounds on the isoperimetric deficit,

$$n|B|\delta(E) \geq \int_E (\lambda_A - \lambda_G)dx, \tag{4.3}$$

$$n|B|\delta(E) \geq \int_{\partial E} (1 - |T|)\, d\mathcal{H}^{n-1}, \tag{4.4}$$

$$n|B|\delta(E) \geq \int_{\partial E} (|T| - (T \cdot \nu_E))\, d\mathcal{H}^{n-1}. \tag{4.5}$$

Let us start examining the first condition. It is well known that equality holds in the arithmetic-geometric mean inequality $\lambda_A \geq \lambda_G$ if and only if there exists $\lambda > 0$ such that $\lambda_k = \lambda$ for $1 \leq k \leq n$. It is easy to turn this characterization of equality cases into a quantitative estimate, and show for example that

$$\lambda_A - \lambda_G \geq \frac{1}{7n^2\lambda_n} \sum_{k=1}^{n} (\lambda_k - \lambda_G)^2, \tag{4.6}$$

see [21, Lemma 2.5]. At the same time [21, Proof of Corollary 2.4] it can be seen that

$$\|\lambda_n\|_{L^1(E)} \leq C(n), \tag{4.7}$$

provided $\delta(E) \leq 1$. Since $\lambda_G = 1$ on E, the combination of (4.3), (4.6) and (4.7) leads to the inequality

$$C(n)\sqrt{\delta(E)} \geq \int_E |\nabla T(x) - \mathrm{Id}|\, dx. \tag{4.8}$$

If we now assume that the Poincaré type inequality

$$|E|^{1/n} \int_E |\nabla u(x)|dx \geq \pi(n) \inf_{c \in \mathbb{R}} \int_E |u - c|, \quad u \in C_c^1(\mathbb{R}^n), \tag{4.9}$$

holds true with a constant $\pi(n)$ depending on the dimension n only, then we can deduce from (4.8) that, up to a translation,

$$C(n)\sqrt{\delta(E)} \geq \pi(n) \int_E |T(x) - x|\, dx. \tag{4.10}$$

We are thus led to consider the following problems:

Problem I. Should we expect the Poincaré type inequality (4.9) to hold on almost optimal sets E, i.e. on sets with $\delta(E) \leq \delta(n)$?

Problem II. Does $\|T(x) - x\|_{L^1(E)}$ control $\alpha(E)$?

Concerning Problem I. Let us first recall that the Poincaré type inequality (4.9) holds true in the form

$$\int_E |\nabla u(x)|dx \geq \pi(E) \inf_{c \in \mathbb{R}} \int_E |u - c|, \quad u \in C_c^1(\mathbb{R}^n), \qquad (4.11)$$

provided we set

$$\pi(E) = \inf \left\{ \frac{\mathcal{H}^{n-1}(E \cap \partial F)}{|F \cap E|} : |F \cap E| \leq \frac{|E|}{2} \right\}.$$

Indeed by the coarea formula

$$\int_E |\nabla u| = \int_{\mathbb{R}} \mathcal{H}^{n-1}(E \cap u^{-1}\{t\})dt.$$

If we now let c be a median for u, *i.e.*

$$|\{u > t\} \cap E| \leq \frac{|E|}{2}, \ t > c, \qquad |\{u < t\} \cap E| \leq \frac{|E|}{2}, \ t < c, \quad (4.12)$$

and we set $v = \max\{u - c, 0\}$, then we find

$$\int_{E \cap \{u > c\}} |\nabla u| = \int_E |\nabla v| = \int_{\mathbb{R}} \mathcal{H}^{n-1}(E \cap v^{-1}\{t\}) \, dt$$

$$= \int_0^\infty \mathcal{H}^{n-1}(E \cap \partial\{v > t\}) \, dt$$

$$\geq \pi(E) \int_0^\infty |E \cap \{v > t\}| \, dt = \pi(E) \int_{E \cap \{u > c\}} |u - c|.$$

By arguing similarly with $v = \max\{c - u, 0\}$ we find that

$$\int_{E \cap \{u < c\}} |\nabla u| \geq \pi(E) \int_{E \cap \{u < c\}} |u - c|,$$

and eventually prove (4.11). How does $\pi(E)$ depend on E? Can we expect that $\pi(E) \geq \pi(n) > 0$ for every E such that $\delta(E) \leq \delta(n)$ and $|E| = |B|$? Unluckily $\pi(E)$ could be zero even for sets E having arbitrarily small isoperimetric deficit $\delta(E)$. For example, $\pi(E) = 0$ whenever E is disconnected. Even if E is connected, $\pi(E)$ can be made arbitrarily small by adding to E sufficiently sharp outward cusps. Both behaviors can safely cohabit with the smallness of the isoperimetric deficit, with the only caveat that the Lebesgue measure of the cusps or of the additional components has to be small in terms of the isoperimetric deficit. This remark suggests that it should be possible to prove a reduction theorem of the following type: there exist constants $C(n)$, $\pi(n)$ and $\delta(n)$ such that for every open set E with $\delta(E) \leq \delta(n)$ there exists $F \subset E$ such that

$$\pi(F) \geq \pi(n), \quad \alpha(E) \leq \alpha(F) + C(n)\delta(E), \quad \delta(F) \leq C(n)\delta(E).$$

Roughly speaking, such a set F is obtained by cutting away from E all the outward cusps and the additional connected components of E, *i.e.* by removing a sort of "maximal bad set" for the Poincaré constant $\pi(E)$, see Figure 4.1.

Concerning Problem II. There is a simple argument leading to the estimate

$$C(n)\sqrt{\int_E |T(x) - x| dx} \geq |E \Delta B|. \tag{4.13}$$

Indeed if we exploit the fact that $T(E) \subset B$ then we find

$$|T(x) - x| \geq \inf\{|z - x| : z \in B\} \geq \varepsilon,$$

whenever $x \in E \setminus B_{1+\varepsilon}$ and $\varepsilon \in (0, 1)$. Thus,

$$\begin{aligned}
|E \Delta B| &= 2|E \setminus B| \\
&\leq 2 \left(|E \setminus B_{1+\varepsilon}| + |B_{1+\varepsilon} \setminus B| \right) \\
&\leq C(n) \left(\frac{1}{\varepsilon} \int_E |T(x) - x| \, dx + n\varepsilon|B| \right),
\end{aligned}$$

and a simple optimization in ε leads to (4.13). As a consequence, if we combine (4.13) with (4.10), the best estimate we can obtain is not sharp, namely, we just find

$$\alpha(E)^4 \leq C(n)\delta(E).$$

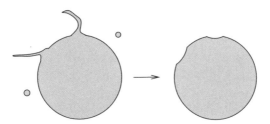

Figure 4.1. Removing a maximal bad set for the constant $\pi(E)$ in the Poincaré inequality. The measure of the region removed is controlled by the size of the isoperimetric deficit.

It is very likely that (4.13) is a non-sharp inequality, and it is natural to conjecture that an inequality like

$$C(n) \int_E |T(x) - x| dx \geq |E \triangle B|, \qquad (4.14)$$

should be true. On combining a reduction theorem for the Poincaré inequality with (4.14) one would immediately prove Theorem 4.1. Although some evidences in favor of (4.14) are presented in [21, Section 1.6], the question of its validity remains open. We can avoid these difficulties by further exploiting the lower bounds on the isoperimetric deficit that are provided by Gromov's argument. We now switch to the proof of Theorem 4.1.

Sketch of the proof of Theorem 4.1. As said, we directly consider the case $K = B$ and assume without loss of generality that E is an open set with C^1-boundary, with $|E| = |B|$ and $\delta(E) \leq \delta(n)$. We let T be the Brenier map between E and B, and then focus on the lower bounds on the isoperimetric deficit (4.10) and (4.4),

$$C(n)|E|\sqrt{\delta(E)} \geq \int_E |\nabla T - \mathrm{Id}|, \qquad (4.15)$$

$$|E|\delta(E) \geq \int_{\partial E} (1 - |T|) d\mathcal{H}^{n-1}, \qquad (4.16)$$

where we recall that $T(E) \subset B$, so that $|T(x)| \leq 1$ on $x \in \overline{E}$.

Step 1. We prove a (possibly degenerate) Poincaré type *trace* inequality on E. We let

$$\tau(E) = \inf \left\{ \frac{\mathcal{H}^{n-1}(E \cap \partial F)}{\mathcal{H}^{n-1}(\partial F \cap \partial E)} : F \subset E, \quad |F| \leq \frac{|E|}{2} \right\}, \qquad (4.17)$$

where F ranges over open sets with piecewise C^1-boundary such that

$$P(F) = \mathcal{H}^{n-1}(E \cap \partial F) + \mathcal{H}^{n-1}(\partial E \cap \partial F). \qquad (4.18)$$

With this definition of $\tau(E)$ we have

$$\int_E |\nabla u| \geq \tau(E) \inf_{c \in \mathbb{R}} \int_{\partial E} |u - c| d\mathcal{H}^{n-1}, \qquad (4.19)$$

for every $u \in C^1(E)$. Indeed if c is a median of u as in (4.12), then the set $F = \{u > t\} = \{x \in E : u(x) > t\}$ is admissible in (4.17) for every $t > c$, so that

$$\mathcal{H}^{n-1}(E \cap \partial \{u > t\}) \geq \tau(E)\mathcal{H}^{n-1}(\partial E \cap \partial \{u > t\}).$$

Let us set $v = \max\{u - c, 0\}$. On the one hand we have that

$$\int_c^\infty \mathcal{H}^{n-1}(E \cap \partial\{u > t\})dt = \int_0^\infty \mathcal{H}^{n-1}(E \cap v^{-1}\{t\})dt$$

$$= \int_E |\nabla v| = \int_{\{u>c\}} |\nabla u|,$$

while on the other hand, since $\{u > t\} = \{x \in E : u(x) > t\}$,

$$\partial E \cap \partial\{u > t\} = \{x \in \partial E : u(x) > t\}.$$

Hence

$$\int_c^\infty \mathcal{H}^{n-1}(\partial E \cap \partial\{u > t\})dt = \int_0^\infty \mathcal{H}^{n-1}(\{x \in \partial E : v(x) > t\})dt$$

$$= \int_{\partial E} v \, d\mathcal{H}^{n-1}$$

$$= \int_{\{x \in \partial E : u(x) > c\}} |u - c| d\mathcal{H}^{n-1}.$$

We have thus proved that

$$\int_{\{x \in E : u(x) > c\}} |\nabla u| \geq \tau(E) \int_{\{x \in \partial E : u(x) > c\}} |u - c| \, d\mathcal{H}^{n-1}.$$

We argue similarly with $v = \max\{c - u, 0\}$ to prove that

$$\int_{\{x \in E : u(x) \leq c\}} |\nabla u| \geq \tau(E) \int_{\{x \in \partial E : u(x) \leq c\}} |u - c| d\mathcal{H}^{n-1},$$

and thus conclude the proof of (4.19).

Step 2. Let us assume for the moment to work with a set E such that $\tau(E) \geq \kappa(n)$, where $\kappa(n)$ is a positive constant depending on the dimension n only. Then by (4.15) we deduce that, up to a translation of E,

$$C(n)\sqrt{\delta(E)} \geq \int_{\partial E} |T(x) - x| d\mathcal{H}^{n-1}. \tag{4.20}$$

If $x \in E$ then we have

$$|1 - |x|| \leq |1 - |T(x)|| + ||T(x)| - |x|| \leq (1 - |T(x)|) + |T(x) - x|,$$

therefore by (4.16) and (4.20) we deduce that

$$\int_{\partial E} |1 - |x|| d\mathcal{H}^{n-1}(x) \leq C(n)(\delta(E) + \sqrt{\delta(E)}) \leq C(n)\sqrt{\delta(E)}, \tag{4.21}$$

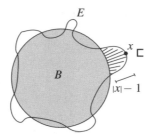

Figure 4.2. When $x \in (\partial E) \setminus B$ then $|x| - 1$ is the length of the segment joining x and $x/|x|$. Thus, up to a corrective Jacobian-like term that is however bounded from below, we have that $\int_{(\partial E) \setminus B} (|x| - 1) d\mathcal{H}^{n-1}$ controls $|E \setminus B|$.

provided $\delta(E) \leq 1$. Eventually we remark that

$$\int_{\partial E} |1 - |x|| \, d\mathcal{H}^{n-1}(x) \geq \int_{(\partial E) \setminus B} (|x| - 1) \, d\mathcal{H}^{n-1}(x)$$
$$\geq c(n)|E \setminus B|, \tag{4.22}$$

see Figure 4.2, and the proof of the theorem is achieved combining (4.21) and (4.22). Therefore we are left to show that we can always reduce to the case that E satisfies $\tau(E) \geq \kappa(n)$ for some positive constant $\kappa(n)$ depending on the dimension n only.

Step 3. We provide a "bad set removal tool", *i.e.* we show that there exists a constant $\kappa(n) \in (0, 1)$ such that if $F \subset E$ and

$$|F| \leq \frac{|E|}{2}, \quad \frac{\mathcal{H}^{n-1}(E \cap \partial F)}{\mathcal{H}^{n-1}(\partial F \cap \partial E)} \leq \kappa(n),$$

then

$$P(E \setminus F) \leq P(E), \tag{4.23}$$
$$|F| \leq C(n)\delta(E)^{n/(n-1)}|E|, \tag{4.24}$$
$$\delta(E \setminus F) \leq C(n)\delta(E), \tag{4.25}$$
$$\alpha(E) \leq C(n)\{\alpha(E \setminus F) + \delta(E)\}. \tag{4.26}$$

Provided $\kappa(n) < 1$, we have that

$$P(E \setminus F) = \mathcal{H}^{n-1}(\partial E \setminus \overline{F}) + \mathcal{H}^{n-1}(E \cap \partial F)$$
$$= \mathcal{H}^{n-1}(\partial E \setminus \overline{F}) + \kappa(n)\mathcal{H}^{n-1}(\partial F \cap \partial E) \leq P(E),$$

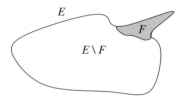

Figure 4.3. When $F \subset E$ so that (4.18) holds true, then $P(E \setminus F) + P(F) = P(E) + 2\mathcal{H}^{n-1}(E \cap \partial F)$.

that is (4.23). To prove (4.24) we first notice that (see Figure 4.3),

$$
\begin{aligned}
P(E) &= P(E \setminus F) + P(F) - 2\mathcal{H}^{n-1}(E \cap \partial F) \\
&\geq P(E \setminus F) + P(F) - 2\kappa(n)\mathcal{H}^{n-1}(\partial F \cap \partial E) \\
&\geq P(E \setminus F) + (1 - 2\kappa(n))P(F),
\end{aligned}
$$

since $P(F) \geq \mathcal{H}^{n-1}(\partial F \cap \partial E)$. Thus, by the isoperimetric inequality

$$
P(E) \geq n|B|^{1/n}\{|E \setminus F|^{(n-1)/n} + (1 - 2\kappa(n))|F|^{(n-1)/n}\}.
$$

If we let $s = |F|/|E|$, subtract $n|E| = n|B|$ and divide by $n|B|$, then we find

$$
\delta(E) \geq (1 - s)^{(n-1)/n} + (1 - 2\kappa(n))s^{(n-1)/n} - 1 = \Psi(s) - 2\kappa(n)s^{(n-1)/n},
$$

where $\Psi : [0, 1] \to \mathbb{R}$ is the concave function

$$
\Psi(s) = (1 - s)^{(n-1)/n} + s^{(n-1)/n} - 1, \quad 0 < s < 1.
$$

Let $\kappa(n)$ be the greatest constant such that

$$
\Psi(s) \geq 3\kappa(n)s^{(n-1)/n}, \quad \forall s \in \left(0, \frac{1}{2}\right).
$$

We have then proved

$$
\delta(E) \geq \kappa(n)s^{(n-1)/n}, \quad i.e. \quad |F| \leq C(n)\delta(E)^{n/(n-1)}|E|,
$$

that is (4.24). To prove (4.25) we notice that by (4.23) and (4.24) we have that

$$
\begin{aligned}
\delta(E \setminus F) &= \frac{P(E \setminus F)}{n|B|^{1/n}|E \setminus F|^{(n-1)/n}} - 1 \\
&\leq (1 - s)^{(n-1)/n} \frac{P(E)}{n|B|^{1/n}|E|^{(n-1)/n}} - 1 \\
&\leq (1 + C(n)\delta(E))\frac{P(E)}{n|B|^{1/n}|E|^{(n-1)/n}} - 1 \leq C(n)\delta(E),
\end{aligned}
$$

provided $\delta(n)$ is small enough. The proof of (4.26) is similar and we omit the details.

Step 4. We now construct a *maximal* bad set in E. More precisely, let us consider the family of sets

$$\Gamma = \left\{ F \subset \mathbb{R}^n : |F| \leq \frac{|E|}{2}, \mathcal{H}^{n-1}(E \cap \partial F) \leq \kappa(n)\mathcal{H}^{n-1}(\partial F \cap \partial E) \right\}.$$

If $\Gamma = \emptyset$ then $\tau(E) \geq \kappa(n)$ and the proof of the theorem is achieved by step two. If otherwise $\Gamma \neq \emptyset$ then we show the existence of $G \subset E$ such that

$$\alpha(E) \leq C(n)\{\alpha(G) + \delta(E)\}, \quad \delta(G) \leq C(n)\delta(E), \tag{4.27}$$
$$\tau(G) \geq \kappa(n). \tag{4.28}$$

To this end we construct a sequence $\{F_h\}_{h \in \mathbb{N}}$ as follows. We let F_1 be any element of Γ. We then let

$$\Gamma_h = \{F \in \Gamma : F_h \subset F\}, \quad s_h = \sup_{F \in \Gamma_h} |F|,$$

and then choose F_{h+1} to be any element of Γ_h such that

$$|F_{h+1}| \geq \frac{|F_h| + s_h}{2}.$$

Then F_h converges increasingly to a set F_∞ such that $|F_\infty| \leq |E|/2$. Moreover by the lower semicontinuity properties of the distributional perimeter,

$$\mathcal{H}^{n-1}(E \cap \partial F_\infty) \leq \liminf_{h \to \infty} \mathcal{H}^{n-1}(E \cap \partial F_h)$$
$$\leq \kappa(n) \liminf_{h \to \infty} \mathcal{H}^{n-1}(\partial F_h \cap \partial E)$$
$$\leq \kappa(n)\mathcal{H}^{n-1}(\partial F_\infty \cap \partial E),$$

where in the last inequality the monotonicity of $\{F_h\}_{h \in \mathbb{N}}$ has been used (since $F_h \subset F_\infty \subset E$ we have that $\partial F_h \cap \partial E \subset \partial F_\infty \cap \partial E$). Thus $F_\infty \in \Gamma$. In fact F_∞ is a maximal element of Γ, in the sense that if $H \subset E \setminus \overline{F_\infty}$ exists so that $F_\infty \cup H \in \Gamma$, then $|H| = 0$. Indeed, since $F_h \subset F_\infty \cup H$ we have $F_\infty \cup H \in \Gamma_h$, therefore

$$s_h \geq |F_\infty \cup H| \geq |F_{h+1} \cup H| = |F_{h+1}| + |H| \geq \frac{s_h + |F_h|}{2} + |H|,$$

i.e.

$$|H| \leq \frac{s_h - |F_h|}{2} \leq |F_{h+1}| - |F_h| \to 0,$$

as $h \to \infty$. Eventually, let us set

$$G = E \setminus \overline{F_\infty}.$$

By step three and by $F_\infty \in \Gamma$ we have (4.27). Let us now prove that, thanks to the maximality of F_∞ in Γ, we have (4.28), *i.e.* $\tau(G) \geq \kappa(n)$. We argue by contradiction and consider the case $\tau(G) < \kappa(n)$. This means that there exists a set H such that

$$|H| \leq \frac{|G|}{2}, \quad \mathcal{H}^{n-1}(G \cap \partial H) \leq \kappa(n)\mathcal{H}^{n-1}(\partial H \cap \partial G). \qquad (4.29)$$

We now prove that $F_\infty \cup H \in \Gamma$, thus contradicting the maximality of F_∞ in Γ (we suggest to keep an eye on Figure 4.4 while reading the following estimates). By (4.29), H is a bad set for G. Hence we can apply step three to H and G to see that

$$|H| \leq C(n)\delta(G)^{n/(n-1)}|G| \leq C(n)\delta(E)^{n/(n-1)}|E|.$$

Hence, since in turn F_∞ is a bad set for E,

$$|F_\infty \cup H| = |H| + |F_\infty| \leq C(n)\delta(E)^{n/(n-1)}|E|.$$

Thus, provided $\delta(n)$ is small enough, we are sure that

$$|F_\infty \cup H| \leq \frac{|E|}{2}. \qquad (4.30)$$

To conclude that $F_\infty \cup H \in \Gamma$ we are left to show that

$$\mathcal{H}^{n-1}(E \cap \partial(F_\infty \cup H)) \leq \kappa(n)\mathcal{H}^{n-1}(\partial(F_\infty \cup H) \cap \partial E). \qquad (4.31)$$

Indeed

$$\mathcal{H}^{n-1}(E \cap \partial(F_\infty \cup H)) = \mathcal{H}^{n-1}(E \cap (\partial F_\infty \setminus \partial H))$$
$$+ \mathcal{H}^{n-1}(E \cap (\partial H \setminus \partial F_\infty)). \qquad (4.32)$$

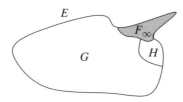

Figure 4.4. Proving the maximality of F_∞.

Since H is a bad set for G,

$$\mathcal{H}^{n-1}(E\cap(\partial H\setminus\partial F_\infty))=\mathcal{H}^{n-1}(G\cap\partial H)\leq\kappa(n)\mathcal{H}^{n-1}(\partial G\cap\partial H)$$
$$\leq\kappa(n)[\mathcal{H}^{n-1}(\partial H\cap\partial E)+\mathcal{H}^{n-1}(E\cap\partial F_\infty\cap\partial H)],$$

therefore from (4.32) we deduce that

$$\mathcal{H}^{n-1}(E\cap\partial(F_\infty\cup H))$$
$$\leq\mathcal{H}^{n-1}(E\cap(\partial F_\infty\setminus\partial H))+\kappa(n)[\mathcal{H}^{n-1}(\partial H\cap\partial E)$$
$$+\mathcal{H}^{n-1}(E\cap\partial F_\infty\cap\partial H)]$$
$$\leq\mathcal{H}^{n-1}(E\cap\partial F_\infty)+\kappa(n)\mathcal{H}^{n-1}(\partial H\cap\partial E)$$
$$\leq\kappa(n)[\mathcal{H}^{n-1}(\partial E\cap\partial F_\infty)+\mathcal{H}^{n-1}(\partial H\cap\partial E)]$$
$$=\kappa(n)\mathcal{H}^{n-1}(\partial(F_\infty\cup H)\cap\partial E),$$

that is (4.31). □

5 Further problems

The methods presented in these notes can be applied in a number of related situations. Without trying to be exhaustive, we discuss here some of these problems.

5.1 Sobolev inequalities

The work of Brezis and Lieb [9] has opened the problem of establishing improvements of Sobolev type inequalities. In the archetypal problem we let $n\geq 2$, $p\in(1,n)$, $p^*=np/(n-p)$, and consider the *Sobolev inequality* on $\mathbb{R}^n$

$$\|\nabla u\|_{L^p(\mathbb{R}^n;\mathbb{R}^n)}\geq S(n,p)\|u\|_{L^{p^*}(\mathbb{R}^n)}. \tag{5.1}$$

By definition, the optimal functions in the Sobolev inequality are the minimizers of the variational problem

$$S(n,p)=\inf\left\{\frac{\|\nabla u\|_{L^p(\mathbb{R}^n;\mathbb{R}^n)}}{\|u\|_{L^{p^*}(\mathbb{R}^n)}}:u\neq 0\right\}. \tag{5.2}$$

It was proved by Aubin [1] and Talenti [50] that minimizers of (5.2) have the form

$$u_{a,r,x_0}(x)=\frac{a}{(1+|r(x-x_0)|^{p/(p-1)})^{(n/p)-1}},\qquad x\in\mathbb{R}^n,$$

for $a \neq 0$, $r > 0$, and $x_0 \in \mathbb{R}^n$. In analogy with Theorem 3.1 and Theorem 4.1, we are concerned with improvements of the Sobolev inequality of the form

$$\|\nabla u\|_{L^p(\mathbb{R}^n;\mathbb{R}^n)} \geq S(n, p)\|u\|_{L^{p^\star}(\mathbb{R}^n)} \left\{1 + \frac{\alpha(u)^{\gamma(n,p)}}{C(n, p)}\right\}, \qquad (5.3)$$

where $\alpha(u)$ is the distance of u from the set of optimal functions $\{u_{a,r,x_0}\}$. In the case $n \geq 3$ and $p = 2$ it was shown by Bianchi and Egnell [5] that (5.3) holds true with

$$\gamma = 2, \quad \alpha(u) = \inf_{a,r,x_0} \frac{\|\nabla u - \nabla u_{a,r,x_0}\|_{L^2(\mathbb{R}^n;\mathbb{R}^n)}}{\|\nabla u\|_{L^2(\mathbb{R}^n;\mathbb{R}^n)}}.$$

This is the best result one can hope to obtain, concerning the exponent γ and the notion of distance involved. The proof of this result is very direct and clean, as it exploits the Hilbert space structure of $W^{1,2}(\mathbb{R}^n)$ to perform a second order Taylor expansion of $\alpha(u)$ for u close to the $(n + 2)$-dimensional manifold $\{u_{a,r,x_0}\}$. When $p \neq 2$ the problem seems to require a different approach. In the limit case $p = 1$ the Sobolev inequality has to be cast on functions of bounded variation, see (1.5) and Remark 1.3. In [29] the symmetrization approach used in the proof of Theorem 3.1 has been adapted to this situation to prove that

$$|Du|(\mathbb{R}^n) \geq n|B|^{1/n}\|u\|_{L^{n/(n-1)}(\mathbb{R}^n)} \left\{1 + \frac{\alpha(u)^2}{C(n)}\right\}, \qquad (5.4)$$

where

$$\alpha(u) = \inf_{a,r,x_0} \left\{ \frac{\int_{\mathbb{R}^n} |u - a\,1_{B(x_0,r)}|^{n/(n-1)}}{\int_{\mathbb{R}^n} |u|^{n/(n-1)}} : \int_{\mathbb{R}^n} |u|^{n/(n-1)} = a^{n/(n-1)}|B|r^n \right\}.$$

By a mass transportation argument in [21, Theorem 4.1] it was also shown that, provided (for simplicity) $u \in C_c^\infty(\mathbb{R}^n)$, $u \geq 0$, one has (5.4) with

$$\alpha(u) = \inf_{t>0} \frac{\int_{\{u>t\}} |\nabla u| + \int_{\mathbb{R}^n\setminus\{u>t\}} |u|^{n/(n-1)}}{\int_{\mathbb{R}^n} |u|^{n/(n-1)}}.$$

This is an analogous result for the case $p = 1$ to the stability theorem in the gradient norm by Bianchi and Egnell. In the case $p \neq 2$, $p > 1$, the only result that is available up to date has been obtained in [12] by a combination of both the symmetrization and the mass transportation methods discussed in these notes. The symmetrization method requires some new ideas expecially in the reduction to n-symmetric functions, due to the scaling invariance of Sobolev minimizers, and it allows to reduce the stability problem to the case of radially symmetric decreasing functions. This case is solved by means of the mass transportation proof of the Sobolev inequality by Cordero-Erausquin, Nazaret and Villani [15] – that takes now the place of Gromov's proof of the Wulff inequality – and thanks to some sharp Sobolev inequalities on domains from [41]. The outcome is (5.3) with

$$\gamma(n, p) = \left(3 + 4p - \frac{3p + 1}{n}\right)^2,$$

$$\alpha(u) = \inf_{a,r,x_0} \left\{ \frac{\|u - u_{a,r,x_0}\|^{p^\star}_{L^{p^\star}(\mathbb{R}^n)}}{\|u\|^{p^\star}_{L^{p^\star}(\mathbb{R}^n)}} : \|u\|_{L^{p^\star}(\mathbb{R}^n)} = \|u_{a,r,x_0}\|_{L^{p^\star}(\mathbb{R}^n)} \right\}.$$

The exponent $\gamma(n, p)$ is clearly non-optimal. Moreover the result of Bianchi and Egnell suggests the possibility of using a stronger notion of α based on the L^p-distance of gradients instead of the $L^{p^\star}$-distance of the functions. Eventually we mention that in the case $p > n$ we have the Sobolev-Morrey inequality

$$|\text{spt}(u)|^{(1/n)-(1/p)} \|\nabla u\|_{L^p(\mathbb{R}^n;\mathbb{R}^n)} \geq C(n, p)\|u\|_{L^\infty(\mathbb{R}^n)}.$$

A quantitative improvement of this inequality was obtained by Cianchi in [11].

5.2 Isoperimetric inequalities in mathematical physics

After the pioneering book by Pólya and Szegö [46], the term "isoperimetric inequality" has been largely used to indicate inequalities expressing the optimality of balls in the volume constrained minimization of various set functions of physical or geometric interest. To show a first class of examples, given $n \geq 2$, $p \in (1, \infty)$ and q such that

$$\begin{cases} q \in [1, p^\star), & \text{if } 1 < p < n, \\ q \in (1, \infty), & \text{if } p \geq n, \end{cases}$$

we consider the set function $\lambda_{p,q}$ defined on open sets with finite measure $\Omega \subset \mathbb{R}^n$ as

$$\lambda_{p,q}(\Omega) = \inf \left\{ \int_\Omega |\nabla u|^p : \int_\Omega |u|^q = 1, \quad u \in W_0^{1,p}(\Omega) \right\}.$$

When $p = q$, the corresponding set function $\lambda_{p,p}$ is the first eigenvalue of the p-Laplacean, while the case $n = 2$, $p = 2$ and $q = 1$ corresponds to the (reciprocal of the) torsional rigidity of Ω. It can be shown that whenever $|\Omega| = |B|$ then we have the *Faber-Krahn inequality*

$$\lambda_{p,q}(\Omega) \geq \lambda_{p,q}(B), \tag{5.5}$$

with equality if and only if Ω is a ball. Another example of physical relevance is that of *p-isocapacitary inequalities*. If Ω is an open set with finite measure and $1 < p < n$ we define the *p-capacity* of Ω as

$$\text{Cap}_p(\Omega) = \inf \left\{ \int_{\mathbb{R}^n} |\nabla u|^p : u \geq 1_\Omega, u \in L^{p^*}(\mathbb{R}^n) \right\}.$$

If $|\Omega| = |B|$ then we have the *p-isocapacitary inequality*

$$\text{Cap}_p(\Omega) \geq \text{Cap}_p(B), \tag{5.6}$$

with equality if and only if Ω is a ball. Eventually, an example of more geometric flavor is that of the Cheeger inequality: given $m \in [1, n/(n-1))$ we define the *m-Cheeger constant* of the open set with finite measure Ω as

$$h_m(\Omega) = \inf \left\{ \frac{P(A)}{|A|^m} : A \subset \Omega \text{ is open} \right\}.$$

Again, if $|\Omega| = |B|$ then we have the *Cheeger inequality*

$$h_m(\Omega) \geq h_m(B), \tag{5.7}$$

with equality if and only if Ω is a ball. Quantitative versions of these isoperimetric inequalities have received some attention, and various cases have been discussed in [3, 42] and [37] concerning the first eigenvalue of the *p*-Laplacean and in [4, 36, 37] concerning the *p*-capacity. A unified approach based on symmetrization theory and covering at once (with non sharp results) all the above mentioned inequalities has been developed in [30]. In the case of the Cheeger inequality an ad-hoc argument leading to a sharp estimate was presented in [22]. The problem of finding sharp forms (in the decay rate of the asymmetry) of the Faber-Krahn and of the isocapacitary inequalities remains open.

5.3   Brunn-Minkowski type inequalities

We have already met the Brunn-Minkowski inequality,

$$|E + F|^{1/n} \geq |E|^{1/n} + |F|^{1/n}. \tag{5.8}$$

When both E and F have positive measure, then equality holds in (5.8) if and only if E and F are (equivalent) to homothetic open convex sets, see [31]. A quantitative version of this statement seems hard to obtain. However, if we further assume that E and F are convex, we can infer a quantitative version of the Brunn-Minkowski inequality from Theorem 4.1. We have already seen that the Brunn-Minkowski inequality implies the Wulff inequality, but the converse is also true when (5.8) is casted on convex sets. Indeed if E and F are bounded, open convex sets then, with the notation introduced in Remark 2.1, we have that

$$f_{E+F}(v) = f_E(v) + f_F(v),$$

for every $v \in S^{n-1}$, and in particular

$$\mathcal{F}_{E+F}(G) = \mathcal{F}_E(G) + \mathcal{F}_F(G), \quad \forall G \subset \mathbb{R}^n.$$

Since $\mathcal{F}_E(E) = n|E|$, $\mathcal{F}_F(F) = n|F|$ and $\mathcal{F}_{E+F}(E+F) = n|E+F|$ we conclude by the Wulff inequality that

$$n|E+F| = \mathcal{F}_{E+F}(E+F) = \mathcal{F}_E(E+F) + \mathcal{F}_F(E+F)$$
$$\geq n|E|^{1/n}|E+F|^{1/n'} + n|F|^{1/n}|E+F|^{1/n'},$$

that is the Brunn-Minkowski inequality. Combining this argument with Theorem 4.1 it is not hard to derive the following quantitative version of the Brunn-Minkowski inequality *restricted to convex sets*.

Theorem 5.1. *If E and F are open bounded convex sets and we set*

$$\sigma(E; F) = \max \left\{ \frac{|F|}{|E|}, \frac{|E|}{|F|} \right\}, \tag{5.9}$$

then we have

$$|E+F|^{1/n} \geq \left(|E|^{1/n} + |F|^{1/n} \right) \left\{ 1 + \frac{\alpha(E; F)^2}{C(n)\sigma(E; F)^{1/n}} \right\}. \tag{5.10}$$

Although is easy to derive (5.10) from Theorem 4.1, we remark that a direct - and, overall, simpler - mass transportation proof of (5.10) can also be obtained, as done in [23]. Other quantitative versions of the Brunn-Minkowski inequality were presented in [17, 32, 33, 47, 48]. There are many other examples of geometric functionals satisfying Brunn-Minkowski type inequalities, for which equality cases have been characterized and for which the question of stability is open: a complete picture of the state of the art can be found in [14]. Moreover Brunn-Minkowski type inequalities are well known to be related to convolution type functional inequalities, see [31]. The quantitative study of these inequalities is widely open.

5.4   Isoperimetric inequalities on manifolds with densities

After the work of Perelman on the Poincaré conjecture, isoperimetric problems in manifolds with density have received an increasing attention, see Morgan [45, Chapter 18] (and the references therein). In the simplest case where the underlying manifold is $\mathbb{R}^n$ we consider a positive density function $F : \mathbb{R}^n \to (0, \infty)$, and study the isoperimetric problems

$$\inf\left\{\int_{\partial E} F(x)d\mathcal{H}^{n-1}(x) : \int_E F(x)dx = m\right\}, \quad m > 0.$$

In the case $F = 1$ this of course reduces to the Euclidean isoperimetric problem. Another non-trivial and very important case where a full characterization of minimizers is available is obtained by looking at the Gaussian density $F(x) = e^{-|x|^2/2}$. In this case minimizers are known to be half-spaces, as proved by probabilistic arguments by Sudakov and Tsirel'son [49] and by Borell [7]. A more geometric proof of this fact can be devised by a symmetrization procedure in the Gauss space, known as the Ehrhard symmetrization [19]. A sharp quantitative Gaussian isoperimetric inequality (in the spirit of Theorem 3.1) has been obtained in [13] by developing in the Gaussian framework the symmetrization tools presented in these notes. Apart from this case, the characterization of minimizers for isoperimetric problems with (suitably simple) densities – and thus, the study of the related stability problems – is open.

ACKNOWLEDGEMENTS. It is a pleasure to thank the organizers of the School "Optimal Transportation, Geometry and Functional Inequalities" for the opportunity to hold these lectures. The author acknowledges the support of the ERC Advanced Grant 2008 *Analytic Techniques for Geometric and Functional Inequalities*.

References

[1] T. AUBIN, *Problèmes isopérimétriques et espaces de Sobolev*, J. Differential Geometry **11** (1976), 573–598.

[2] F. BERNSTEIN, *Über die isoperimetriche Eigenschaft des Kreises auf der Kugeloberflache und in der Ebene*, Math. Ann. **60** (1905), 117–136.

[3] T. BHATTACHARYA, *Some observations on the first eigenvalue of the p-Laplacian and its connections with asymmetry*, Electron. J. Differential Equations **35** (2001), 15.

[4] T. BHATTACHARYA and A. WEITSMAN, *Bounds for capacities in terms of asymmetry*, Rev. Mat. Iberoamericana **12** (1996), 593–639.

[5] G. BIANCHI and H. EGNELL, *A note on the Sobolev inequality*, J. Funct. Anal. **100** (1991), 18–24.

[6] T. BONNESEN, *Über die isoperimetrische Defizit ebener Figuren*, Math. Ann. **91** (1924), 252–268.

[7] C. BORELL, *The Brunn-Minkowski inequality in the Gauss space*, Invent. Math. **30** (1975), 207–216.

[8] Y. BRENIER, *Polar factorization and monotone rearrangement of vector-valued functions*, Comm. Pure Appl. Math. **44** (1991), 375–417.

[9] H. BREZIS and E. H. LIEB, *Sobolev inequalities with remainder terms*, J. Funct. Anal. **62** (1985), 73–86.

[10] A. CIANCHI, *A quantitative Sobolev inequality in BV*, J. Funct. Anal. **237** (2006), 466–481.

[11] A. CIANCHI, *Sharp Sobolev-Morrey inequalities and the distance from extremals*, Trans. Amer. Math. Soc. **360** (2008), 4335–4347.

[12] A. CIANCHI, N. FUSCO, F. MAGGI and A. PRATELLI, *The sharp Sobolev inequality in quantitative form*, Journal of the European Mathematical Society **11** (2009), 1105–1139.

[13] A. CIANCHI, N. FUSCO, F. MAGGI and A. PRATELLI, *On the isoperimetric deficit in the Gauss space*, Accepted on the American Journal of Mathematics.

[14] A. COLESANTI, *Brunn-Minkowski inequalities for variational functionals and related problems*, Adv. Math. **194** (2005), 105–140.

[15] D. CORDERO-ERAUSQUIN, B. NAZARET and C. VILLANI, *A mass-transportation approach to sharp Sobolev and Gagliardo-Nirenberg inequalities*, Adv. Math. **182** (2004), 307–332.

[16] A. DINGHAS, *Über einen geometrischen Satz von Wulff für die Gleichgewichtsform von Kristallen*, Z. Kristallogr., Mineral. Petrogr. (German) **105**, (1944).

[17] V. I. DISKANT, *Stability of the solution of a Minkowski equation*, Sibirsk. Mat. Ž. (Russian) **14** (1973), 669–673, 696.

[18] E. DE GIORGI, *Sulla proprietá isoperimetrica dell'ipersfera, nella classe degli insiemi aventi frontiera orientata di misura finita*, Atti Accad. Naz. Lincei. Mem. Cl. Sci. Fis. Mat. Nat. Sez. I (Italian) **5** (1958), 33–44.

[19] A. EHRHARD, *Symétrisation dans l'espace de Gauss*, Math. Scand. (French) **53** (1983), 281–301.

[20] L. ESPOSITO, N. FUSCO and C. TROMBETTI, *A quantitative version of the isoperimetric inequality: the anisotropic case*, Ann. Sc. Norm. Super. Pisa Cl. Sci. (5) **4** (2005), 619–651.

[21] A. FIGALLI, F. MAGGI and A. PRATELLI, *A mass transportation approach to quantitative isoperimetric inequalities*, submitted paper.

[22] A. FIGALLI, F. MAGGI and A. PRATELLI, *A note on Cheeger sets*, Proc. Amer. Math. Soc. **137** (2009), 2057–2062.

[23] A. FIGALLI, F. MAGGI and A. PRATELLI, *A refined Brunn-Minkowski inequality for convex sets*, Ann. Inst. H. Poincaré Anal. Non Linéaire **26** (2009), 2511–2519.

[24] I. FONSECA and S. MÜLLER, *A uniqueness proof for the Wulff theorem*, Proc. Roy. Soc. Edinburgh Sect. A **119** (1991), 125–136.

[25] B. FUGLEDE, *Lower estimate of the isoperimetric deficit of convex domains in R^n in terms of asymmetry* Geom. Dedicata **47** (1993), 41–48.

[26] B. FUGLEDE, *Stability in the isoperimetric problem for convex or nearly spherical domains in $\mathbb{R}^n$*, Trans. Amer. Math. Soc. **314** (1989), 619–638.

[27] N. FUSCO, *The classical isoperimetric theorem*, Rend. Accad. Sci. Fis. Mat. Napoli **71** (2004), 63–107.

[28] N. FUSCO, F. MAGGI and A. PRATELLI, *The sharp quantitative isoperimetric inequality*, Ann. of Math. **168** (2008), 941–980.

[29] N. FUSCO, F. MAGGI and A. PRATELLI, *The sharp quantitative Sobolev inequality for functions of bounded variation*, J. Funct. Anal. **244** (2007), 315–341.

[30] N. FUSCO, F. MAGGI and A. PRATELLI, *Stability estimates for certain Faber-Krahn, isocapacitary and Cheeger inequalities*, Ann. Sc. Norm. Super. Pisa Cl. Sci. (5) **8** (2009), 51–71.

[31] R. J. GARDNER, *The Brunn-Minkowski inequality*, Bull. Amer. Math. Soc. (N.S.) **39** (2002), 355–405.

[32] H. GROEMER, *On the Brunn-Minkowski theorem*, Geom. Dedicata **27** (1988), 357–371.

[33] H. GROEMER, *On an inequality of Minkowski for mixed volumes*, Geom. Dedicata **33** (1990), 117–122.

[34] H. HADWIGER and D. OHMANN, *Brunn-Minkowskischer Satz und Isoperimetrie*, Math. Zeit. **66** (1956), 1–8.

[35] R. R. HALL, *A quantitative isoperimetric inequality in n-dimensional space*, J. Reine Angew. Math. **428** (1992), 161–176.

[36] R. R. HALL, W. K. HAYMAN and A. W. WEITSMAN, *On asymmetry and capacity*, J. d'Analyse Math. **56** (1991), 87–123.

[37] W. HANSEN and N. NADIRASHVILI, *Isoperimetric inequalities in potential theory*, Proceedings from the International Conference on Potential Theory (Amersfoort, 1991). Potential Anal. **3** (1994), 1–14.

[38] F. JOHN, *An inequality for convex bodies*, Univ. Kentucky Research Club Bull. **8** (1942), 8–11.

[39] H. KNOTHE, *Contributions to the theory of convex bodies*, Michigan Math. J. **4** (1957) 39–52.

[40] F. MAGGI, *Some methods for studying stability in isoperimetric type problems*, Bull. Amer. Math. Soc. **45** (2008), 367-408.

[41] F. MAGGI and C. VILLANI, *Balls have the worst best Sobolev inequalities*, J. Geom. Anal. **15** (2005), 83–121.

[42] A. MELAS, *The stability of some eigenvalue estimates*, J. Differential Geom. **36** (1992), 19–33.

[43] R. J. MCCANN, *Existence and uniqueness of monotone measure-preserving maps*, Duke Math. J. **80** (1995), 309–323.

[44] V. D. MILMAN and G. SCHECHTMAN, *Asymptotic theory of finite-dimensional normed spaces*, with an appendix by M. Gromov, "Lecture Notes in Mathematics", 1200, Springer-Verlag, Berlin, 1986, viii+156.

[45] F. MORGAN, "Geometric measure theory", a beginner's guide. Fourth edition, Elsevier/Academic Press, Amsterdam, 2009, viii+ 249.

[46] G. PÓLYA, and G. SZEGÖ, "Isoperimetric Inequalities in Mathematical Physics", Annals of Mathematics Studies, n. 27, Princeton University Press, Princeton, NJ, 1951.

[47] I. Z. RUZSA, *The Brunn-Minkowski inequality and nonconvex sets*, (English summary) Geom. Dedicata **67** (1997), 337–348.

[48] R. SCHNEIDER, *On the general Brunn-Minkowski theorem*, Beiträge Algebra Geom. **34** (1993), 1–8.

[49] V. N. SUDAKOV and B. S. TSIREL'SON, *Extremal properties of half-spaces for spherically invariant measures*, J. Soviet Math. (1978), 9-18.

[50] G. TALENTI, *Best constant in Sobolev inequality*, Ann. Mat. Pura Appl. **110** (1976), 353–372.

[51] J. E. TAYLOR, *Crystalline variational problems*, Bull. Amer. Math. Soc. **84** (1978), 568–588.

CRM Series
Publications by the Ennio De Giorgi Mathematical Research Center Pisa

The Ennio De Giorgi Mathematical Research Center in Pisa, Italy, was established in 2001 and organizes research periods focusing on specific fields of current interest, including pure mathematics as well as applications in the natural and social sciences like physics, biology, finance and economics. The CRM series publishes volumes originating from these research periods, thus advancing particular areas of mathematics and their application to problems in the industrial and technological arena.

Published volumes

1. Matematica, cultura e società 2004 (2005). ISBN 88-7642-158-0
2. Matematica, cultura e società 2005 (2006). ISBN 88-7642-188-2
3. M. GIAQUINTA, D. MUCCI, *Maps into Manifolds and Currents: Area and $W^{1,2}$-, $W^{1/2}$-, BV-Energies*, 2006. ISBN 88-7642-200-5
4. U. ZANNIER (editor), *Diophantine Geometry.* Proceedings, 2005 (2007). ISBN 978-88-7642-206-5
5. G. MÉTIVIER, *Para-Differential Calculus and Applications to the Cauchy Problem for Nonlinear Systems*, 2008. ISBN 978-88-7642-329-1
6. F. GUERRA, N. ROBOTTI, *Ettore Majorana. Aspects of his Scientific and Academic Activity*, 2008. ISBN 978-88-7642-331-4
7. Y. CENSOR, M. JIANG, A. K. LOUISR (editors), *Mathematical Methods in Biomedical Imaging and Intensity-Modulated Radiation Therapy (IMRT)*, 2008. ISBN 978-88-7642-314-7
8. M. ERICSSON, S. MONTANGERO (editors), *Quantum Information and Many Body Quantum systems.* Proceedings, 2007 (2008). ISBN 978-88-7642-307-9
9. M. NOVAGA, G. ORLANDI (editors), *Singularities in Nonlinear Evolution Phenomena and Applications.* Proceedings, 2008 (2009). ISBN 978-88-7642-343-7

Matematica, cultura e società 2006 (2009). ISBN 88-7642-315-4

10. H. HOSNI, F. MONTAGNA (editors), *Probability, Uncertainty and Rationality*, 2010. ISBN 978-88-7642-347-5

11. L. AMBROSIO (editor), *Optimal Transportation, Geometry and Functional Inequalities*, 2010. ISBN 978-88-7642-373-4

Volumes published earlier

Dynamical Systems. Proceedings, 2002 (2003)
  Part I: *Hamiltonian Systems and Celestial Mechanics*.
ISBN 978-88-7642-259-1
  Part II: *Topological, Geometrical and Ergodic Properties of Dynamics*.
ISBN 978-88-7642-260-1

Matematica, cultura e società 2003 (2004). ISBN 88-7642-129-7

Ricordando Franco Conti, 2004. ISBN 88-7642-137-8

N.V. KRYLOV, *Probabilistic Methods of Investigating Interior Smoothness of Harmonic Functions Associated with Degenerate Elliptic Operators*, 2004. ISBN 978-88-7642-261-1

Phase Space Analysis of Partial Differential Equations. Proceedings, vol. I, 2004 (2005). ISBN 978-88-7642-263-1

Phase Space Analysis of Partial Differential Equations. Proceedings, vol. II, 2004 (2005). ISBN 978-88-7642-263-1

Fotocomposizione "CompoMat" Loc. Braccone, 02040 Configni (RI) Italy
Finito di stampare nel mese di febbraio 2010
dalla CSR, Via di Pietralata 157, 00158 Roma